T0305882

Nanotechnology for Smart Concrete

Emerging Materials and Technologies

Series Editor
Boris I. Kharissov

Biomass for Bioenergy and Biomaterials
Nidhi Adlakha, Rakesh Bhatnagar, and Syed Shams Yazdani

Energy Storage and Conversion Devices: Supercapacitors, Batteries, and Hydroelectric Cell
Anurag Gaur, A.L. Sharma, and Anil Arya

Nanomaterials for Water Treatment and Remediation
Srabanti Ghosh, Aziz Habibi-Yangjeh, Swati Sharma, and Ashok Kumar Nadda

2D Materials for Surface Plasmon Resonance-Based Sensors
Sanjeev Kumar Raghuwanshi, Santosh Kumar, and Yadvendra Singh

Functional Nanomaterials for Regenerative Tissue Medicines
Mariappan Rajan

Uncertainty Quantification of Stochastic Defects in Materials
Liu Chu

Recycling of Plastics, Metals, and Their Composites
R.A. Ilyas, S.M. Sapuan, and Emin Bayraktar

Viral and Antiviral Nanomaterials
Synthesis, Properties, Characterization, and Application
Devarajan Than gadurai, Saher Islam, Charles Oluwaseun Adetunji

Drug Delivery using Nanomaterials
Yasser Shahzad, Syed A.A. Rizvi, Abid Mehmood Yousaf and Talib Hussain

Nanomaterials for Environmental Applications
Mohamed Abou El-Fetouh Barakat and Rajeev Kumar

Nanotechnology for Smart Concrete
Ghasan Fahim Huseien, Nur Hafizah A. Khalid, and Jahangir Mirza

Nanomaterials in the Battle Against Pathogens and Disease Vectors
Kaushik Pal and Tean Zaheer

MXene-Based Photocatalysts: Fabrication and Applications
Zuzeng Qin, Tongming Su, and Hongbing Ji

For more information about this series, please visit: https://www.routledge.com/Emerging-Materials-and-Technologies/book-series/CRCEMT

Nanotechnology for Smart Concrete

Ghasan Fahim Huseien, Nur Hafizah A. Khalid,
Jahangir Mirza

CRC Press
Taylor & Francis Group
Boca Raton London New York

CRC Press is an imprint of the
Taylor & Francis Group, an **informa** business

First edition published 2022
by CRC Press
6000 Broken Sound Parkway NW, Suite 300, Boca Raton, FL 33487-2742

and by CRC Press
2 Park Square, Milton Park, Abingdon, Oxon, OX14 4RN

© 2022 Taylor & Francis Group, LLC

CRC Press is an imprint of Taylor & Francis Group, LLC

Library of Congress Cataloging-in-Publication Data
Names: Huseien, Ghasan Fahim, author. | Nur Hafizah A. Khalid, author. |
Mirza, Jahangir, 1944- author.
Title: Nanotechnology for smart concrete / Ghasan Fahim Huseien, Nur
Hafizah A. Khalid, Jahangir Mirza.
Description: First edition. | Boca Raton : CRC Press, 2022. | Series:
Emerging materials and technologies | Includes bibliographical
references and index. | Summary: "Nanomaterials have been shown to markedly improve the mechanical properties of concrete, as well as to reduce the porosity and enhance the durability of concrete. Nanotechnology for Smart Concrete discusses the advantages and applications of nanomaterials in concrete including high-strength performance, microstructural improvement, self-healing, energy storage, and coatings. It analyzes the linkage of concrete materials with nanomaterials and nanostructures. It also discusses applications of nanomaterials in the concrete industry, including energy storage in green buildings, anti-corrosive coatings, and inhibiting pathogens and viruses"—Provided by publisher.
Identifiers: LCCN 2021047664 (print) | LCCN 2021047665 (ebook) | ISBN
9781032051277 (hbk) | ISBN 9781032051284 (pbk) | ISBN 9781003196143 (ebk)
Subjects: LCSH: Concrete—Additives. | Nanostructured materials.
Classification: LCC TP884.A3 H87 2022 (print) | LCC TP884.A3 (ebook) |
DDC 666/.893—dc23/eng/20211122
LC record available at https://lccn.loc.gov/2021047664
LC ebook record available at https://lccn.loc.gov/2021047665

ISBN: 978-1-032-05127-7 (hbk)
ISBN: 978-1-032-05128-4 (pbk)
ISBN: 978-1-003-19614-3 (ebk)

DOI: 10.1201/9781003196143

Typeset in Times
by codeMantra

Contents

Preface...xi
Authors.. xiii

Chapter 1   Nanotechnology and Nanomaterials: An Introduction 1

      1.1   Introduction 1
      1.2   Definition of Nanoscience and Nanotechnology.....................2
      1.3   Nanoscience and Nanotechnology 2
      1.4   Nanoparticles Preparation Methods 4
      1.5   Applications of Nanomaterials in Construction.....................6
      References 7

Chapter 2   Applications of Nanomaterials and Nanotechnology in the
                Construction Industry... 11

      2.1   Introduction 11
      2.2   Significance of Nanotechnology in Construction Engineering...11
      2.3   Nanotechnologies for Concrete 13
      2.4   Nanotechnology in the Cement Industry.............................. 14
      2.5   Asphalt... 15
      2.6   Brick 16
      2.7   Steel 16
      References 17

Chapter 3   Nanomaterial-Based Cement Concrete: Engineering Properties
                and Durability Performance.. 21

      3.1   Introduction 21
      3.2   Nanomaterial-Modified Cement Binder................................22
      3.3   Fresh Properties..23
      3.4   Strength Performance 24
      3.5   Microstructure Properties...25
      3.6   Durability Properties 27
      3.7   Summary 29
      References 30

Chapter 4   Sustainability of Nanomaterial-Based Self-Healing Concrete...........35

      4.1   Introduction 35
      4.2   Sustainability of Smart Concrete 36
      4.3   Lifecycle Analysis of Self-Healing Concrete 36
      4.4   Production of Nanomaterials...37

4.5   Production of Nanoconcrete 38
4.6   Significance of Nanomaterials as Self-Healer........................38
4.7   Nano-Silica-Based Self-Healing Concrete 39
4.8   Nanoalumina-Based Self-Healing Concrete 40
4.9   Carbon Nanotube-Based Self-Healing Concrete....................40
4.10   Titanium Oxide-Based Self-Healing Concrete.......................41
4.11   Nanokaolin and Nanoclay-Based Self-Healing Concrete........42
4.12   Nanoiron-Based Self-Healing Concrete 43
4.13   Economy of Nanomaterial-Based Self-Healing Concretes 43
4.14   Safety Features of Nanomaterial-Based Concretes.................44
4.15   Summary 44
References 45

Chapter 5   Engineering Properties of High-Performance Alkali-Activated
            Mortars 49

5.1   Introduction 49
5.2   Workability Performance 51
       5.2.1   Flowability..51
       5.2.2   Viscosity 52
       5.2.3   Setting Time 53
       5.2.4   Bulk Density..53
5.3   Mechanical Properties..55
       5.3.1   Compressive Strength..55
       5.3.2   Splitting Tensile Strength 56
       5.3.3   Flexural Strength..57
       5.3.4   Modulus of Elasticity 57
       5.3.5   Relationship between CS, STS, and FS....................58
       5.3.6   Statistical Data Analysis 58
5.4   Microstructure Properties...60
       5.4.1   XRD Patterns of AAMs..60
       5.4.2   FESEM Analyses 61
       5.4.3   EDX Analyses 62
       5.4.4   FTIR Spectra..65
       5.4.5   TGA and DTG Curves 66
5.5   Summary 67
References 68

Chapter 6   Effect of Nanomaterials on Durability Properties of Free
            Cement Mortars..73

6.1   Introduction 73
6.2   Water Absorption..74
6.3   Drying Shrinkage 76
6.4   Carbonation Depth 77
6.5   Abrasion Resistance 79

6.6   Resistance to Freeze-Thawing Cycles 81
6.7   Acid Attack Resistance... 86
6.8   Resistance to Elevated Temperatures 90
6.9   Summary 96
References 97

Chapter 7   Nanoparticle-Based Phase Change Materials for Sustained
               Thermal Energy Storage in Concrete.. 99

7.1   Introduction 99
7.2   Overview of Energy Storage 100
        7.2.1   Energy Storage Methods 101
        7.2.2   Latent Heat Storage (LHS).. 102
        7.2.3   Methods for Latent Heat of Fusion (LHF) and
                Melting Temperature Measurement 102
        7.2.4   TES System 102
7.3   Phase Change Materials 103
        7.3.1   Phase Change Materials Classification 103
        7.3.2   Organic Materials... 104
        7.3.3   Inorganic Materials 105
        7.3.4   Eutectics Materials 106
        7.3.5   Hygroscopic Materials 106
        7.3.6   Solid to Solid Phase Transforming Materials 107
7.4   Criteria of PCMs to Be Used for TES.................................. 107
7.5   Criteria of TES Based on PCMs for Building Applications... 107
7.6   Benefits of TES Based on PCMs for Buildings.................... 108
7.7   Technology, Development, and Encapsulation 108
7.8   Nanomaterial-Based PCMs 109
        7.8.1   Preparation Methods 109
        7.8.2   Interfacial Polymerization.. 110
        7.8.3   Emulsion Polymerization Method (EP) 110
        7.8.4   Miniemulsion Polymerization (MEP) Method........ 111
        7.8.5   In Situ Polymerization (ISP) Method...................... 113
        7.8.6   Sol–Gel Method 113
7.9   PCM Incorporation Procedures in Concrete 116
        7.9.1   Immersion Technique.. 116
        7.9.2   Impregnation Technique.. 117
        7.9.3   Direct Mixing Technique 117
7.10   Effects of CPM on Concrete Properties 118
        7.10.1 Fresh Properties.. 118
        7.10.2 Mechanical Characteristics 119
        7.10.3 Durability Properties... 120
7.11   Stability of PCM in Concrete 120
7.12   Effects of Nanomaterials on Enhancement TES of PCM 121
7.13   Applications of NE-PCMs in Concrete 123
7.14   Concrete Thermal Energy Storage with NE-PCMs.............. 125

7.15 Environmental Effects 125
7.16 Energy and Sustainability 126
7.17 Suggestions for Future Works 127
7.18 Summary 127
References 128

Chapter 8 Concrete Coatings: Applications of Nanomaterials and
Nanotechnology 137
8.1 Introduction 137
8.2 Concrete Durability 139
8.3 Coating Technology in Concrete 141
8.4 Nanotechnology and Nanoparticles 143
8.5 Nanoparticles and Nanomaterials 144
8.6 Definition of Nanotechnology in Concrete 145
8.7 Nanomaterial-Based Concretes 145
8.8 Production of Nanoconcrete 146
8.9 Nanomaterial-Based Concrete Surface Coating 147
     8.9.1 Polymer Nanocomposite Coatings 147
     8.9.2 Silane–Clay Nanocomposite Coatings 148
     8.9.3 SiO_2 and/or TiO_2 Nanoparticle Coating Concrete 149
8.10 Summary 151
References 151

Chapter 9 Nanomaterial-Based Concrete Antivirus Surface 159
9.1 Introduction 159
9.2 Nanomaterial-Based Antimicrobial Coatings 161
9.3 Types of Nanomaterials for Antimicrobial Coatings 163
     9.3.1 Copper (Cu) 163
     9.3.2 Silver (Ag) 165
     9.3.3 Silica (SiO_2) 168
     9.3.4 Titanium (TiO_2) 168
     9.3.5 Zinc Oxide (ZnO) 170
9.4 Summary 172
References 172

Chapter 10 Nanomaterials: Environmental Health and Safety Considerations ... 181
10.1 Introduction 181
10.2 Safety Considerations 182
10.3 Potential Risks and Concerns 184
10.4 Risk of Nanomaterials 185
10.5 Toxicity of Nanomaterials 186
10.6 Using Nanomaterials Safely 187
10.7 Mitigation of Public and Environmental Health Impacts 188

Contents   ix

10.8 Risk Communication and Technological Impact
        Assessment 189
10.9 Environmental Implications 190
10.10 Summary 191
References 192

Chapter 11 Sustainability and Environmental Benefits of Concrete with
             Nanomaterials 195

11.1 Introduction 195
11.2 Life Cycle Assessment 197
11.3 Sustainability of Nanomaterial-Based Concrete ... 200
11.4 Merits and Demerits of Nanomaterials for Concrete ... 201
11.5 Economy of Nanomaterial-Based Concretes 201
11.6 Environmental Suitability of Nanomaterial-Based
        Concretes 203
References 203

Appendix: Questions and Answers 205
Index 209

Preface

The concrete industry is known to consume enormous amount of natural resources extracted from the planet. Cement is the major reason for this large consumption. Scientists and researchers have therefore focused their attention to replace the conventional ingredients of concrete structure with a new material to minimize the use of natural resources and increase its durability. Nanomaterials have emerged as the promising new material in the 21st century. Contemporary concrete production techniques place more emphasis on the inclusion of nanomaterials that offer enhanced features irrespective of whether it is fresh or cured.

Nanotechnology for Smart Concrete is the first edition, comprising 11 chapters under the series Emerging Materials and Technologies. Each of these chapters discusses one of the applications of nanomaterials in the concrete industry. Inclusion of nanomaterials in concrete matrix not only improves the strength properties, but it was found to enhance the durability and sustainability of the proposed concrete.

The first chapter deals with the general introduction of nanomaterials, definition of nanoscience and nanotechnology, nanoparticles preparation methods, and applications of nanomaterials in the construction industry. Several applications of nanoparticle materials in the construction and concrete industries are widely discussed in Chapter 2. In Chapter 3, we focus on the benefits of nanomaterials for strength and microstructure enhancement. The effect of nanomaterial type, particle size, and content on strength development at early and later age has been discussed.

The benefits of nanomaterials for smart and high-performance concrete such as self-healing, free cement concrete, high durable concrete and coating applications are presented in Chapters 3–9. Topics regarding safety, toxicity, environmental and health impact, sustainability benefits as well as future utilization are presented in Chapters 10 and 11.

The utilization of nanomaterials for sustainable concrete production and in the construction industry whose growth knows no boundaries has been studied. Mounting evidence of worldwide interest suffices the need to produce a collective anthology of a wide variety of nanomaterials available today.

Authors

Dr. Ghasan Fahim Huseien is a research fellow at Department of the Build Environment, School of Design and Environment, National University of Singapore, Singapore. He received his PhD degree from University Technology Malaysia in 2017. He is involved in "Applied R & D research on evaluation and application of various types of materials to repair cracks, damaged surfaces in concrete" and "Applied R & D and utilization of industrial and agriculture wastes (to reduce cost, energy and environmental problems) for the existing and future construction of concrete structures, both in the laboratory and in the field". He has authored and co-authored more than 45 technical report and significant publications, 2 books, and 8 book chapters. He is peer reviewer for several international journals as well as Master's and PhD students. He is a member of Concrete Society of Malaysia and American Concrete Institute.

Dr. Ghasan Fahim Huseien has over 5 years of Applied Research and Development (R&D) as well as up to 10 years of experience in manufacturing smart materials for sustainable building and smart cities. He has expertise in Advanced Sustainable Construction Materials covering Civil Engineering, Environmental Sciences and Engineering, Chemistry, Earth Sciences, Geology, and Architecture departments. He has authored and co-authored +75 publications and technical reports, 4 books, and 17 book chapters and participated in +30 national and international conferences/workshops.

His past experience in projects includes the application of nanotechnology in construction and building materials, self-healing technology, and geopolymers as sustainable and eco-friendly repair materials in the construction industry.

Dr. Nur Hafizah Abd Khalid is a senior lecturer at School of Civil Engineering, Universiti Teknologi Malaysia (UTM) and a research member of Construction Material Research Group (CMRG). Previously, she has been appointed as Council Member of Concrete Society Malaysia (CSM) for 2 years. She completed all her degrees in Universiti Teknologi Malaysia (UTM) and completed her PhD in 2016. Along her career and research journey, she has received many awards such as Young Women Scientist 2014 (Korea), Talented Young Scientist 2018–2019 (China), and the ASEAN-ROK Next Innovator 2020 (Korea). Her research interest focuses on Advanced Concrete Materials, Green and Environmental-Friendly Concrete, Nanomaterials, Characterization of Materials, and Chemical Dynamics of Materials.

Her biggest interest is in recycling wastes into value-added concrete material to promote low-embodied energy construction material and bridge the gap between research innovation and industrial application.

Professor Dr. Jahangir Mirza (speaks six languages: English, French, German, Hindi, Punjabi, and Urdu) has over 35 years of Applied Research and Development (R&D) as well as teaching experience. He has expertise in Advanced Sustainable Construction Materials covering Civil Engineering, Environmental Sciences and Engineering, Chemistry, Earth Sciences, Geology, and Architecture departments. His past and current experiences are as follows: (i) 1985 till July 2019: Senior Scientist, Research Institute of Hydro-Québec (IREQ), Montreal, Canada "Applied R&D research on evaluation and application of various types of materials to repair cracks, joints, water-stops, damaged surfaces in concrete hydraulic- and infra-structures" and "Applied R&D and utilization of industrial and agriculture wastes (to reduce cost, energy and environmental problems) for the existing and future construction of concrete structures, both in the laboratory and in the field". (ii) September 2018 till present: Visiting Research Professor, Environmental Engineering program, School of Engineering, University of Guelph, Ontario, Canada: "Conduct research on reuse of incinerations". (iii) May 2014 till June 2016: Professor, UTM Construction Research Centre, Faculty of Civil Engineering, Universiti Teknologi Malaysia, Johor Bahru, Malaysia: Applied R & D: "utilization of industrial, agricultural and natural wastes (to reduce cost, energy and environmental problems) in concrete"; "evaluation and application of various materials to repair cracks, joints, water-stops, damaged surfaces in concrete structures"; and "development of geopolymer (cement-free) mortars and concrete using all kinds of wastes for new structures and as repair material". Supervisor and co-supervisor of MSc and PhD Civil Engineering students, taught courses to BSc, MSc, and PhD students; "Advanced Concrete Technology", "Advanced Construction Materials and Environmental Issue", write research proposals and publications. (iv) 1992–2000: Adjunct Professor, Department of Civil Engineering and Applied Mechanics, McGill University, Montreal, Canada: conducted a course for graduate and post-graduate Civil Engineering students; supervised and co-supervised Master's and PhD Civil Engineering students in their applied research. (v) 1981–1984: Research Scientist, Canada Cement Lafarge, Ltd.: conducted and supervised various R&D and troubleshooting projects related to cement, mortar, concrete problems (chemical and physical characterization and their testing, alkali-aggregate reaction, corrosion in reinforced concrete, slag, fly ashes, silica fumes, grouts, admixtures, gypsum, and other related construction materials). Other accomplishments and activities are as follows: (i) authored and co-authored +200 publications and technical reports, 2 books, and 15 book chapters. (ii) Participated in +40 national and international conferences/workshops (+30 cities, +16 countries). (iii) Supervised and co-supervised MSc, PhD Civil Engineering students in Canada, Pakistan, and Malaysia. (iv) Recipient of +30 national and international awards/prizes/honours from Canada, Indonesia,

Korea, Malaysia, Pakistan, Thailand, United Kingdom, and United States (include six gold medals and one silver medal), TWO BEST PAPER awards from International Conferences: Structural Faults & Repairs, United Kingdom (1993) and first International Conference on Durability of Buildings and Infrastructures (DuraBI2018), Miri, Malaysia (2018). (v) Member Editorial Board of Q1 Journal – *International Journal of Construction and Building Materials*, published in United Kingdom, since 1993. (vi) Conducted presentations, upon invitation, in Canada, France, Germany, Jordan, Korea, Pakistan, Malaysia, Romania, Thailand, Turkey, Singapore, South Africa, United Arab Emirates, and United States.

1 Nanotechnology and Nanomaterials
An Introduction

1.1 INTRODUCTION

An outbursting field of science, nanotechnology has tremendous potentials to affect positively in a limitless variety of applications. First identified in 1958, and developed through numerous stages since then, nanotechnology is the combination and development of knowledge gained from a wide variety of other scientific fields such as engineering, chemistry, biology, physics, medicine, and informatics. Through its ongoing developments, the field of nanotechnology can produce nanomaterials with different additional functions, properties, and characteristics which make them unique and enhance in a range of issues faced by humanity today, including health, environment, construction, and electronics [1].

Nanotechnology can be defined as the manipulation of shape and structure of materials at the nanoscale in order to design, characterize, and produce useful structures, devices, and systems. The nanoscale refers to the objects between 1 and 100 nm in size, where 1 nm is equal to 1×10^{-9} m. Many challenges exist when working at such a small scale, with advanced imaging techniques which is a prerequisite to study and improve materials' behaviour and to design and manufacture them with nanoscale particle size (such as very fine powders, liquids, or solids). These are known as nanoparticles [2,3].

Selected as one of the top ten targeted applications of nanotechnology to ameliorate some of the developing world's most significant issues, construction and architecture industries stand to be substantially enhanced by the employment of nanomaterials [4]. Although already being used within these contexts [4,5], the future of nanotechnology in these industries is predicted to further augment many of its applications. Among these expected improvements are the building material properties themselves by making them stronger, more durable, and lighter [6,7]; by introducing novel collateral functions, such as self-heating, anti-fogging, and energy-saving coatings [8,9]; and which are the key components for the maintenance of instruments, such as sensors that detect and report structural health [10]. However, despite the new technologies can provide advantages, it is worth noting that an insufficient emphasis should be placed on the risk assessment of their intended use, the fallout can be severe. Recent one such example of this was the deliberate and widespread use of supposedly beneficial substance dichlorodiphenyltrichloroethane (DDT), a chemical that was released to control malaria and various waterborne diseases, but instead proved to be carcinogenic to humans, toxic to numerous bird species, and hazardous

DOI: 10.1201/9781003196143-1

to the environment [11]. This illustrates the importance of a proactive and meticulous approach to risk assessments for new technologies, without which, devastating impacts to ecosystems and human health cannot be prevented.

1.2   DEFINITION OF NANOSCIENCE AND NANOTECHNOLOGY

Although nanoscience and nanotechnology share the same prefix 'nano', from the Greek meaning 'dwarf', the two words have distinctly different meanings: nanoscience refers to the study and observation of molecules and structures at the nanoscale, whereas nanotechnology is the practical application of this knowledge in technological devices or systems [12,13]. Another distinction worth mentioning is that while nanotechnology has only been in development since the 1950s, nanoscience can be traced back to Democritus in 5th century BCE Greece, when the question initially arose as to whether matter is finite or continuous, and the concept of small indivisible and indestructible particles – now known as atoms – was first mentioned. To facilitate visualization of the sizes involved at nanoscale, the following examples are worth considering: the thickness of a single human hair is around 60,000 nm, and the radius of the double helix structure of DNA is 1 nm [14].

Through converting the theory and knowledge gained from nanoscience into practical technological applications, nanotechnology positions itself as one of the most promising technologies of the 21st century. The US National Nanotechnology Initiative (NNI) has a definition of nanotechnology that implies the existence of two conditions: "a science, engineering, and technology conducted at the nanoscale (1–100 nm), where unique phenomena enable novel applications in a wide range of fields, from chemistry, physics, and biology, to medicine, engineering, and electronics" [15]. These conditions are as follows: first, the scale at which this technology is used must be at the nanometre scale; and secondly, these innovations must be novel in their exploitation of unique working facets at the nanoscale [16,17].

Summarily, while nanoscience is the overlapping of physics, materials science, and biology, combined directly to the observation of matter at molecular and atomic scales, nanotechnology applies this theoretical knowledge to manipulate, design, and fabricate matter at the nanometre scale. Although a number of reports exist which provide details of their history, no one paper covers effectively the entire history. As a result, it is crucial that a wide range of sources should be analysed and summarized in the following sections to ensure a complete foundational awareness for the development of these fields [18].

1.3   NANOSCIENCE AND NANOTECHNOLOGY

Science itself represents the most powerful tool not only to comprehend the surrounding environment to humankind but also to effect change within that space. At the inception of scientific approaches, the primary goal was to understand nature and the physical world. With time, this focus shifted from theoretical knowledge to practical applications, changing from a phase of pure discovery to one of invention.

In this regard, science and technology are interconnected, whereby science is the unearthing of new information, and technology is the transformation of these insights into new opportunities, in the form of useful tools, materials, and objects. By stating this, it is logical then that there is a relationship between nanoscience and nanotechnology, with both referring more specifically to the scale on which these discoveries and inventions are made: between atoms, at tenths of nanometres, and molecules, of many nanometres [19]. Specifically, nanotechnology applies to any approaches which produce devices and designs finer than 1 mm and larger than 1 nm [20]. Furthermore, while considering that all matter is made up of atoms, it could be logical to also infer that nanoscience and nanotechnology cover all other branches of science and technology. However, a much more accurate definition of these fields is achieved when specifically considering the intrinsic properties of nanoscale objects and the advantages that they provide. Nanoscience and nanotechnology are positioned within the broad interdisciplinary area which comprise of polymers and metal particles, nanoelectronics, supramolecular and colloid chemistry, nanostructured materials, biochemistry, and biology.

Although the recent 5 years have demonstrated a surge in the nanoscience and nanotechnology fields, nanoparticles already saw their frequent use in the fabrication of stained glass, whereby ruby-red Au and lemon-yellow Ag particles were used for colour [16]. The earliest and most prominent example of this is from the 4th century where coloured-glass lined bronze artefact was found known as the Roman Lycurgus Cup. Despite the fact that the synthesis of nanoparticles has technically been carried out for centuries, there still exists a further need for their development which will only increase their applications more crucial for medical, computing, and sensor applications. Progress in the field of nanotechnology substantially depends on these developments, as the ability of nanoparticles to manufacture and assemble into desired shapes is presently the limiting factor to be achieved.

For the performance of basic investigations or experimentation, the manipulation of imaging nanoscale approaches, such as atomic force microscopy (AFM) and transmission electron microscopy (TEM), plays a crucial role. With the use of these techniques, nanoscience and nanotechnology fields can focus on the generation of synthetic methods and surface analytical tools, fabrication of materials and structures at the sub-100 nm scale; identification and quantification of the consequences miniaturization can have, both chemically and physically; and invention and subsequent employment of discovered properties for the usefulness of innovative materials, devices, and systems. Given this, it is perhaps no surprise that these are the fields responsible for, and best equipped to handle, developing innovations such as nanoparticles, nanostructured materials, nanoporous materials, nanopigments, nanotubes, nanoimprinting, and quantum dots; the applications of which have already led to significant advances in materials science [16,21].

As is common with relatively new developing fields, nanoscience and nanotechnology produce novel and innovative results with significant potential [22] as well as they produce disappointing results. It is widely agreed upon within the scientific community that a greater degree of control over the molecular arrangement of matter should be attained to expand the possible properties with which matter that could be imbued would expand dramatically – in ways that even our present science is unable to accurately predict [23]. In ways, the very words nanoscience and nanotechnology themselves can inspire intense emotional responses; the scientific and civilian communities alike seem to expect that whatever influence these fields have, positive or negative, will likely be extreme. However, there are many arguments which assert that nanoscience and nanotechnology could contribute in the advancement of a number of large-scale issues facing humanity such as food, health, pollution, and energy. Despite such a widespread awareness of the fields, no one universally accepted the definition which exists for them. It is unlikely that there will ever be and even when that is achieved, other issues will surface.

Given that nanoscience is conducted at the nanometre scale, it is influenced by quantum mechanical effects (most notably of which, the quantum size effect); this illustrates unequivocally how the field must be interdisciplinary in nature, positioned in synergy between physics, chemistry, materials science, and biology. In this manner, when advancements are made within nanoscience, these are shared as a central point in the fundamental research of the aforementioned other areas, and similarly into technology. Presently, microtechnology sees greater importance than the nanotechnology; however, with time this is certain to change, given its revolutionary potential. The more mature scientific fields are also theorized to play a significant role in this, although whether as lead or as support is yet unknown – largely these circumstances will depend upon emerging opportunities, the scientists themselves, and how adaptable and imaginative they can be.

Nanotechnology is not geared towards working at continually diminishing dimensions but rather towards taking advantage of the peculiar behaviours exhibited naturally by materials, specifically at the nanoscale. Although nanoparticles themselves are not new, occurring already at this point with and without human intervention, a number of recent developments in the technology are used to visualize and manipulate them have directly sparked a proliferation of both new materials and applications for their use, in fields from medicine to car manufacturing. Utilizing these novel techniques to fabricate nanostructures, which were previously only theoretical, forms objects that have superior properties than their raw materials. This means that the nanostructures can outperform other forms of the same material in conductivity (heat or electricity); biological properties; capacity to act as chemical catalysts by changing their surface area or their reactivity; physical and mechanical properties; magnetic properties; and even improved optical characteristics, such as reflecting light or changing colour when their structure is altered. One particular example is concrete: as its surface area per mass is increased with the introduction of nanoparticles, its reaction performance is improved because of a greater area of contact with surrounding materials.

1.4   NANOPARTICLES PREPARATION METHODS

In the pursuit of understanding and manipulating matter at the nanoscale, nanotechnology is an emerging field of science that capitalizes on the unique phenomena exhibited by materials at that scale. Nanoscale science, engineering, and technology intersect within this area of study, combined with the tools and techniques to provide imaging, dimensioning, visualizing, and fabricating matters at nanoscale.

Two predominant design approaches exist in the field of nanotechnology: top-down and bottom-up. The former refers to minimizing the size of existing structures to the greatest possible extent while retaining their original functions and features, without control on the atomic level – as frequently seen in electronics – or, alternatively, through deconstruction into smaller constituent parts. The latter is also referred to Ref. [23] as molecular manufacturing or molecular nanotechnology and utilizes assembly or self-assembly to engineer materials from raw atoms or molecular components (Figure 1.1). Despite the top-down approach being favoured in a number of modern technologies, it is still worth mentioning the potential within the alternate approach, particularly for inciting breakthroughs in the industries, such as biotechnology, information technology, national security, materials and manufacturing, electronics, and medicine and healthcare.

There exist numerous seismic developments in physics, chemistry, and biology which utilized and proved Feynman's theories on the manipulation of matter at

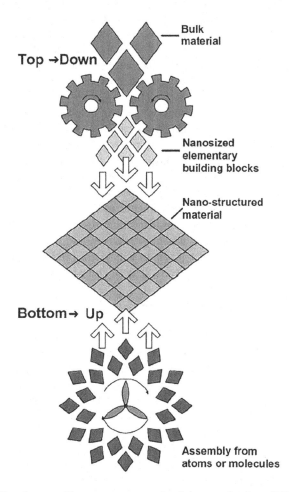

FIGURE 1.1 Top-down and bottom-up approaches in nanotechnology [28].

the nanoscale in late 50s, when he first gave his widely renowned lecture 'There's Plenty of Room at the Bottom' [1]. Additionally, despite once being a novel term, 'nanotechnology' has now carried a meaning – albeit one that varies across different scientific fields and countries and is often misused to signify any minor technology. However, now this word is even more frequently understood to indicate the understanding, restructuring, and manipulation of matters at the nanoscale to fabricate materials with distinctly novel characteristics and properties [24].

Despite extraordinary potential for possible applications of nanotechnology in the construction and architecture industry, at present the rate of advances in improving building materials has been scattered and unsteady [25]. Recently, thanks to the advancement at nanoscale, the latest cutting-edge information has been revealed about concrete in particular, a portion of which even goes as far as to invalidate previous approaches and understandings. However, the market currently fails to reflect and exploit these discoveries from a marketing point of view and proved largely unsuccessful. A number of these significant advances relate to cementing materials as a whole [26], with many defining principles of their characteristics being rewritten, such as origins of cement cohesion, hydration, concrete interfaces, degradation mechanisms, and structure and mechanical properties of the predominant hydrated phases. Notably, another author also has made efforts to summarize and compile the history of nanotechnology applications within construction industry [27].

The NNI gives concrete-based materials as examples for areas on which nanotechnology stands to have a noticeably large impact. The Presidential Commission on Nanotechnology drew parallels between the potential effects of nanotechnology on humankind to those of the Industrial Revolution. The Commission's report detailed nine 'grand challenges' to the world where nanotechnology had the most substantial potential to affect changes, and one of these challenges was to have a safe and affordable transportation. In simpler terms, the concept is to imitate nature with engineering modifications to molecular structures to affect the bulk qualities of the material; however, in practice, it is rarely this simple. This bottom-up approach should facilitate the ongoing development of long-lasting and high-performance systems and products to integrate nanotechnology in a sustainable manner.

1.5 APPLICATIONS OF NANOMATERIALS IN CONSTRUCTION

When carbon nanotubes (CNTs) and nano-sized SiO_2 (or Fe_2O_3) particles are added to concrete mixtures, they improve dramatically a number of their mechanical properties. This is significant because the concrete has the greatest annual production quantities compared to other construction materials [6,29]. The addition of CNTs improved conventional cement adhesion characteristics and also enhanced the resultant composite's durability and toughness by reinforcing its mechanical strength. Furthermore, it is now possible to effectively prevent crack propagation through the addition of 1 wt% CNTs which act as a nucleating agent [30,31]. Other additives, with beneficial properties, include silica and iron oxide nanoparticles; when added at 3–10 wt%, they can reinforce concrete properties by acting as filling agents [6,7,32].

Another widely used construction material, 'steel' presents its own challenges in its resistance to corrosion, ductility, formability, and strength. A number of these issues can be addressed successfully through the addition of metal nanoparticles [33]. One such excellent example of this is copper, which when added as nanoparticles can reduce the surface roughness of any steel part, in combination with improved weldability and anti-corrosion behaviour [13,33].

The potential benefits nanotechnology can impart to glass are also noteworthy, for example, TiO_2 and SiO_2 nanoparticles are already being used at present to coat windows and impart advantageous properties. The reactive oxygen species (ROS) generated photochemically by a TiO_2 coating on glass triggered by artificial or natural light effectively and efficiently removes bacterial dirt and films present on the surface, essentially creating self-cleaning windows automatically [9,34]. The TiO_2 coating also makes the glass easier to clean, by reducing the contact angle between its surface and water droplets and anti-fogging, as a result of its light-excited super hydrophilic characteristics [8]. As far as the benefits of nanosized silica layers are concerned, when placed in contact between two glass panels, they provide significant fireproofing protection [35].

Apart from building materials, there are a number of other areas within construction industry where nanomaterials can play a vital role as useful products. When silver nanoparticles (nAg) are used as paint additives, they impart biocidal action to the paint, thanks to its antimicrobial activity [36]. Similar to the benefits it provides as a glass coating, TiO_2 coatings act as a solar-powered anti-fouling agent on walls, pavements, paths, and roofs which also aid to keep those surfaces clean [5]. TiO_2 solar cells sensitized to dye and silicon-based photovoltaic panels can now also be used as 'energy coatings' to generate power from the sunlight which are sufficiently flexible to be applied on roofs and windows. Academicians have also recently asserted that the conversion efficiency of numerous fuel and solar cells could drastically be enhanced by the inclusion of CNTs, C_{60} fullerenes, and CdSe quantum dots, to the point where partial non-utility power generation for a household was achieved [37].

REFERENCES

1. Jian, Z., et al., Nasicon-structured materials for energy storage. *Advanced Materials*, 2017. **29**(20): p. 1601925.
2. Khandve, P., Nanotechnology for building material. *International Journal of Basic and Applied Research*, 2014. **4**: pp. 146–151.
3. Rana, A.K., et al., Significance of nanotechnology in construction engineering. *International Journal of Recent Trends in Engineering*, 2009. **1**(4): p. 46.
4. Lee, J., S. Mahendra, and P. Alvarez, Potential environmental and human health impacts of nanomaterials used in the construction industry, in Bittnar Z., Bartos P.J.M., Němeček J., Šmilauer V., Zeman J. (eds) *Nanotechnology in Construction 3.* 2009, Springer: Berlin, Heidelberg, pp. 1–14.
5. Zhu, W., P.J. Bartos, and A. Porro, Application of nanotechnology in construction. *Materials and Structures*, 2004. **37**(9): pp. 649–658.
6. Sobolev, K. and M.F. Gutiérrez, How nanotechnology can change the concrete world: part two of a two-part series. *American Ceramic Society Bulletin*, 2005. **84**(11): pp. 16–19.

7. Li, G., Properties of high-volume fly ash concrete incorporating nano-SiO$_2$. *Cement and Concrete Research*, 2004. **34**(6): pp. 1043–1049.

8. Kontos, A., et al., Superhydrophilicity and photocatalytic property of nanocrystalline titania sol–gel films. *Thin Solid Films*, 2007. **515**(18): pp. 7370–7375.

9. Irie, H., K. Sunada, and K. Hashimoto, Recent developments in TiO$_2$ photocatalysis: novel applications to interior ecology materials and energy saving systems. *Electrochemistry*, 2004. **72**(12): pp. 807–812.

10. Saafi, M. and P. Romine, Nano-and microtechnology. *Concrete International*, 2005. **27**(12): pp. 28–34.

11. Turusov, V., V. Rakitsky, and L. Tomatis, Dichlorodiphenyltrichloroethane (DDT): ubiquity, persistence, and risks. *Environmental Health Perspectives*, 2002. **110**(2): pp. 125–128.

12. Shah, K.W., Nanosynthesis techniques of silica-coated nanostructures. In George Z. Kyzas and Athanasios C. Mitropoulos (eds) *Novel Nanomaterials-Synthesis and Applications*. 2018. Volume 1, pp. 91–109, InTechOpen.

13. Shah, K.W. and G.F. Huseien, Biomimetic self-healing cementitious construction materials for smart buildings. *Biomimetics*, 2020. **5**(4): p. 47.

14. Gnach, A., et al., Upconverting nanoparticles: assessing the toxicity. *Chemical Society Reviews*, 2015. **44**(6): pp. 1561–1584.

15. Hulla, J., S. Sahu, and A. Hayes, Nanotechnology: history and future. *Human & Experimental Toxicology*, 2015. **34**(12): pp. 1318–1321.

16. Bayda, S., et al., The history of nanoscience and nanotechnology: from chemical–physical applications to nanomedicine. *Molecules*, 2020. **25**(1): p. 112.

17. Bhushan, B., Introduction to nanotechnology, in *Springer Handbook of Nanotechnology*. 2017, Springer: Berlin, Heidelberg, pp. 1–19.

18. Satalkar, P., B.S. Elger, and D.M. Shaw, Defining nano, nanotechnology and nanomedicine: why should it matter? *Science and Engineering Ethics*, 2016. **22**(5): pp. 1255–1276.

19. López-Lorente, Á.I. and M. Valcárcel, Analytical nanoscience and nanotechnology, in *Comprehensive Analytical Chemistry*. 2014, Elsevier, pp. 3–35.

20. Shukla, A.K. and S. Iravani, *Green Synthesis, Characterization and Applications of Nanoparticles*. 2018, Elsevier, pp. 1–26.

21. Whitesides, G.M., Nanoscience, nanotechnology, and chemistry. *Small*, 2005. **1**(2): pp. 172–179.

22. Capek, I., *Nanocomposite Structures and Dispersions*, Vol. 23. 2019, Elsevier, pp. 1–76.

23. Drexler, K.E., Nanotechnology: from Feynman to funding. *Bulletin of Science, Technology & Society*, 2004. **24**(1): pp. 21–27.

24. Hulteen, J., A general template-based method for the preparation of nanomaterials. *Journal of Materials Chemistry*, 1997. **7**(7): pp. 1075–1087.

25. Garboczi, E., Concrete nanoscience and nanotechnology: definitions and applications, in Bittnar Z., Bartos P.J.M., Němeček J., Šmilauer V., Zeman J. (eds) *Nanotechnology in Construction 3*. 2009, Springer: Berlin, Heidelberg, pp. 81–88.

26. Lee, J., S. Mahendra, and P.J. Alvarez, Nanomaterials in the construction industry: a review of their applications and environmental health and safety considerations. *ACS Nano*, 2010. **4**(7): pp. 3580–3590.

27. Muzenski, S., et al., Towards ultrahigh performance concrete produced with aluminum oxide nanofibers and reduced quantities of silica fume. *Nanomaterials*, 2020. **10**(11): p. 2291.

28. Sanchez, F. and K. Sobolev, Nanotechnology in concrete–a review. *Construction and Building Materials*, 2010. **24**(11): pp. 2060–2071.

29. Lau, D., et al., Nano-engineering of construction materials using molecular dynamics simulations: prospects and challenges. *Composites Part B: Engineering*, 2018. **143**: pp. 282–291.

30. Saez de Ibarra, Y., et al., Atomic force microscopy and nanoindentation of cement pastes with nanotube dispersions. *Physica Status Solidi (a)*, 2006. **203**(6): pp. 1076–1081.
31. Mohamed, H.A., Applications of nanomaterials in architectural design. *Resourceedings*, 2020. **2**(1): pp. 197–206.
32. Li, H., et al., Microstructure of cement mortar with nano-particles. *Composites Part B: Engineering*, 2004. **35**(2): pp. 185–189.
33. Spitzmiller, M., S. Mahendra, and R. Damoiseaux, Safety issues relating to nanomaterials for construction applications, in Fernando Pacheco-Torgal, Maria Vittoria Diamanti, Ali Nazari, and Claes Goran-Granqvist (eds) *Nanotechnology in Eco-Efficient Construction*. 2013, Volume 1, Elsevier, Woodhead Publishing: Netherlands, pp. 127–158.
34. Paz, Y., et al., Photooxidative self-cleaning transparent titanium dioxide films on glass. *Journal of Materials Research*, 1995. **10**(11): pp. 2842–2848.
35. Mann, S., Nanoforum report: nanotechnology and construction. nanoforum. org. 2006.
36. Kumar, A., et al., Silver-nanoparticle-embedded antimicrobial paints based on vegetable oil. *Nature Materials*, 2008. **7**(3): pp. 236–241.
37. Girishkumar, G., et al., Single-wall carbon nanotube-based proton exchange membrane assembly for hydrogen fuel cells. *Langmuir*, 2005. **21**(18): pp. 8487–8494.

2 Applications of Nanomaterials and Nanotechnology in the Construction Industry

2.1 INTRODUCTION

While the construction industry is experiencing the importance of nanotechnology revolution in particular, many other areas are also being explored from diverse science and engineering to commercial sectors. The improvements in efficacy brought about by changes made in physical and chemical properties at the nanoscale are remarkable. These changes can only be done at the nanoscale in areas, such as mechanical strength, optical sensitivity, thermal and electrical conductivity, and photocatalytic potential, and can see the benefits. These, in turn, facilitate new applications for nanotechnology in catalysts, electrical storage devices, advanced mechanical materials, and sensors [1].

Manufactured nanomaterials (MNMs) have a range of advantages when it comes to their use in construction industry. These include in providing materials which possess superior structural properties, paints and coatings with unique functions, and high-resolution detailed sensing and actuating systems. A range of updated potential applications for MNMs within construction can be seen in Figure 2.1.

2.2 SIGNIFICANCE OF NANOTECHNOLOGY IN CONSTRUCTION ENGINEERING

Nanotechnology presents many opportunities to design materials with properties which never existed previously for use in a very broad range of areas. The nanoparticles are perfectly organized and aligned atoms. This means that substantial alterations occur to the properties of any given material when their dimensions are reduced from macro to micro and then to nanosize [2,3]. Once this size is achieved, the unique chemical and physical properties of nanoparticles then permit the creation of systems with remarkably high sensitivity, functional density, catalytic effects, great surface areas, special surface effects, and high strain resistance [4]. The production of these nanoparticles influences the cosmetics, construction, electronics, manufacturing, and medical industries extraordinarily alike, with the new nanotechnologies finding a variety of new applications [1]. In 2015, the Allied Market Research (AMR) report valued the nanomaterials market

DOI: 10.1201/9781003196143-2

FIGURE 2.1   Applications of manufactured nanomaterials in the construction industry.

at $14,741.6 million and were expected to reach $55,016 million by 2022 [2] – an anticipated increase of nearly four times. In the same year, AMR reported the Nanotechnology Consumer Products Inventory (CPI) aggregated a list of all nano-technological products from across 32 different countries, detailing 1814 products in total produced by 6222 companies. Of these products, the nanomaterial which was found to be used most frequently was silver and the industries that contained the largest quantity of products included health, medicine, manufacturing, elec-tronics, and construction materials [5–8]. The examples of discovered applications for nanoparticles in the field of construction are shown in Table 2.1.

Recent figures from the EPA USA state that the total quantity of built space in the United States will increase to 430 billion square feet from the present 300 billion square feet (generating a significant source of solid waste) in the coming 30 years [9]. Evidently, it is crucial to adjust current approaches to account for this through the utilization of novel practices on a wide scale which might reduce the consumption of natural resources to a sustainable level. Nanotechnology is equipped with the capacity to completely and fundamentally change the con-struction industry to focus on safeguarding the environment and competing with innovations while allowing them to be financially advantageous [10]. This can be achieved by improving the materials' performance which would reduce the raw materials' quantity required for any given task and consequentially reduce the energy consumption of their production. Over all, the use of nanoparticles provides

TABLE 2.1

Potential Applications of Nanomaterials in the Construction Industry

Type of Nanomaterials	Area	Applications
Silica	Concrete	Enhance the mechanical properties
		Improve the hydration process
Titanium	Concrete	Improve the hydration process
		Self-cleaning concrete
		Enhance the strength and durable performance
Carbon nanotubes	Concrete	Engineering properties enhancement
		Crack prevention
Aluminium oxide	Asphalt	Serviceability increment
Clay	Brick	Increase the compressive strength
		Increase the surface roughness
Iron oxide	Concrete	Enhance the strength and abrasion properties
Copper	Steel	Corrosion resistance
Zycosoil	Asphalt	Enhance the compaction and fatigue life

a more economical, efficient, and safer approach when creating construction materials. This would not only be cheaper initially but also be cheaper in terms of costs incurred during the entire life cycle of the nanomaterials by improving their properties [11]. In this regard, lower density and higher strength nanomaterials enhance and reinforce materials' durability, efficiency, and performance, in particular, per unit raw material consumed [10].

2.3 NANOTECHNOLOGIES FOR CONCRETE

Heavily influenced by its nanoproperties, concrete is a commonly used construction material which stands to have its performance significantly improved through the addition of nanoscale materials. As previously stated, it is widely accepted that the addition of carbon nanotubes (CNTs) up to 1 wt% into Portland cement and other cements can improve their mechanical properties substantially (increased both the compressive and flexural strengths) [12]. However, among the types of CNTs which have been investigated so far, oxidized multi-walled nanotubes have demonstrated superior performance in both compressive and flexural strengths of subsequent concretes compared to the conventional samples. Silica also finds its use as a beneficial cement additive, for example, when amorphous nano-silica is employed, it disperses to increase segregation resistance in self-compacting concretes. As explored by Du et al. [13], nano-SiO_2 added to cement-based materials led to improve durability, as it increased the control over the degradation of calcium-silicate-hydrate reaction and blocked the water penetration. The compressive strength of concretes containing large volumes of fly ash could also be greatly improved by adding nano-SiO_2, as the pores between the ash and cement particles could be filled by the addition of nanoparticles.

2.4   NANOTECHNOLOGY IN THE CEMENT INDUSTRY

Cement is one of the most commonly and widely used construction materials [14]. China is the largest producer of cement and manufactured 2 billion tonnes in 2011. Total worldwide production for that year was 3400 million tonnes. After China, India produced 210 million tonnes and the United States 68 million tonnes of cement and seems relatively small quantities.

The increasing pressure for high-performance and low-environmental impact construction materials may already have led to exploit new classes of materials. However, concrete and cement-based materials are still the most widely used. This is despite its manufacture and production being one of the most sizeable contributors of CO_2 emissions worldwide accounts for between 5% and 6% of all human sources of CO_2 per annum [15]. The role that nanotechnology can play in changing this is vital, particularly in terms of most effective applications, and exceptionally so for concrete. Without intervention from nanotechnology, cement-based materials have poor mechanical properties, are highly susceptible to chemicals, and highly permeable to water – all of which reduce their durability and lifespan. However, with the consideration and appropriate incorporation of nanoparticles, such as $nSiO_2$, ZnO_2, Al_2O_3, TiO_2, CNTs, nano-clays, and nanofibres of carbon, those issues can be resolved. Not only that, they provide other benefits, such as improving those materials, among others, poor crack resistance, low tensile strength, extended curing times, high permeability, and low ductility.

While concrete now regularly sees nanoparticles added for its performance, it has been, in fact, in use for a long time as a construction material. A large proportion of the concrete contains Portland cement-based binders, water, and both coarse and fine aggregates. To form the binders, Portland clinker is ground together to fine consistency with gypsum and various mineral powders, including limestone, pozzolan (typically volcanic ash), granulated blast-furnace slag, and fly ash (a waste product from coal-burning power plants) [16,17]. Despite already being advanced so far, it is crucial to continue to improve upon these alterations to the traditional cement binders, as the longevity and durability of concrete surfaces and structures continuously face harsh weather exposures. When a concrete is used for known particular applications, its properties can be adjusted further to produce a material that is specialized for whatever function it is to perform. This could be done by using chemical admixtures like air-entraining agents and superplasticizers in the concrete mixture to modify its properties [18]. Given that the durability of the final concrete mixture is heavily influenced by interface connections between the aggregates, voids, and cement paste [19], it is logical to infer that nanomaterials will play a crucial role in the ongoing development of concrete, particularly those with enhanced durability and strength properties [20].

As mentioned before, there are two main additives – titanium dioxide (TiO_2), and silica (SiO_2) – which are presently used in concrete mixes. Recent studies have shown that when silica is present instead as nanoparticles, called nano-silica ($nSiO_2$), the particle packing is dramatically improved in the concrete [21,22]. Behaving as a nano filler in the cement to surround the calcium-silicate-hydrate (Ca-Si-H) particles, the $nSiO_2$ also serves as a strong binding agent and improves the cement and aggregate cohesion. Nano-silica also reduces the potential concrete degradation risk over time by reducing its porosity, thus lowering the probability of water or other elements to penetrate into the surface of structures [22]. Furthermore, it not only affects the setting time and dormant period, reducing them effectively, but also increases the early strength of cement as a result of increased rate of cement hydration. These effects can be observed in Figure 2.1 which shows scanning electron microscope (SEM) micrographs of a typical cement paste alongside the nano-silica-modified cement paste. The other aforementioned additive, titanium dioxide, is more recently being used in concrete as nanoparticles (nTiO$_2$) and is already being mass-produced, thanks to its stability, photocatalytic, and anticorrosive characteristics [23–25]. The high surface area of nTiO$_2$ results in unique photocatalytic behaviour and acts as active self-sustaining functions which concrete can perform. All powered with its photocatalytic action on the surface, the concrete becomes self-disinfecting, self-cleaning, and can actively remove environmental pollutants, breaking them down into water and CO_2.

On the cutting edge of novel concrete additives, CNTs have greatly enhanced durability and improved mechanical properties which make the concrete impenetrable to water and salts [26,27]. This is a result of high surface area of CNTs and remarkably high mechanical properties. Shah et al. [27] demonstrated that CNTs behaved as significantly more efficient fillers within the concrete pore voids and provided other benefits to its properties.

2.5 ASPHALT

With an ability to carry high volumes of traffic loads, asphalt concrete (asphalt) is a commonly used composite material and is invaluable for how frequently it is used on surfaces of driveways, road, and airport runways. However, despite being employed so intensively for surface covering, it has a very poor resistance against high and low temperatures alike, often melting or cracking at extreme temperatures [28]. Adjustments can be made by using nanoparticles in asphalt to improve aforementioned limitations, thus enhancing its ductility and elasticity to compensate for its poor thermal stability. The additives most frequently being used at present include styrene-butadiene-styrene (SBS), polyethylene (PE), and styrene-butadiene-rubber (SBR) [28,29]. Recently, a new asphalt additive – aluminium oxide nanoparticles (nAl$_2$O$_3$), has shown promising results in trials. It was found that a resistance to extremely high temperatures could be achieved by incorporating 5% proportion of nAl$_2$O$_3$ in an asphalt mix [30]. In the future, this form of additive appears to be well-positioned to find a place as an alternative to the list of regularly used asphalt modifiers. The integration of nanoparticles into asphalt has irrefutably and substantially improved its serviceability as a construction material [30,31].

A number of recent studies have shown to enhance the issue of moisture damage in asphalt pavements over a time. It is a significant problem as the moisture can infiltrate into the pavement structure and cause failure to the hot mix asphalt layers, leading to loss of durability, strength, and stiffness [32]. One of these studies centred on the effects of zycosoil, when added as an anti-strip agent to asphalt concrete, and how it influenced its resulting properties [33,34]. Findings demonstrated that the aggregate coverage of zycosoil generated an increase in serviceable fatigue life and reduced fatigue wear. This is due to the resulting smaller number of air voids and increased filler which improved compaction and enabled the aggregate surface modification [33–35].

2.6 BRICK

Brick is construction material which has seen use for centuries. Brick currently constitutes 50%–80% clay and the rest is sand and other granular materials. Combined, compressed, and fired under high temperatures, the resulting material possesses good compressive strength, making bricks ideally suited for application in domestic construction [36,37]. However, simply because the strength is good does not mean it cannot be improved further as Niroumand et al. [38] tried to discover when they experimented using nano clays in earth bricks, measuring resulting changes to compressive strengths [38]. Their results illustrated a vast potential for nanoparticles to influence materials' performance. By including 5% nano clay, the brick's relative strength increased up to 4.8 times compared with conventional bricks [38]. Widely regarded as layered mineral silicates of nanoparticles, nano clays appear to be the most sustainable additive to work with thus far in bricks, particularly when considering their service life. Also, additives react within the internal structure, and preservative coatings are applied externally to the bricks to affect their properties. In particular, linseed oil, silane, siloxane, and alkosiloxane containing 1%–1.5% nano-silica were experimentally tested by Stefanidou and Karazou [39]. They found that preservative coatings were the most effective to date to preserve bricks and increased their longevity and durability by significantly reducing their permeability.

2.7 STEEL

Possessing the highest strength-to-weight ratio of commonly used construction materials, steel finds extensive use as building material because of its durability, fire resistance, and comparatively sustainable as well as due to its recyclable nature not causing inherent degradation to reuse. However, despite their reigning status, new bars, known as micro-composite, multi-structural formable steel (MMFX) have started to gain preference, possessing superior durability and corrosion resistance to plain steel and with nanoparticle-infused paint coatings reinforcing them within concrete structures [10]. Other studies have also illustrated that the influence on improving the steel microstructure can minimize the negative effects of hydrogen embrittlement and the inter-granular cementite phase [10]. The uneven surfaces frequently found in typical steels generate stresses where the forces are concentrated on a small area and will begin to form fatigue-induced cracks before they propagate along a surface. This reduces the surface unevenness by adding nanoparticle additives which will subsequently decrease the risk of cracking [11].

Another form of coating investigated by Hegazy et al. involved the addition of colloidal copper nanoparticles to anti-corrosion paints designed for use on steel [40]. They fabricated them by using the chemical reduction of copper (II) chloride ($CuCl_2$), as a colloidal dispersion solution at various concentrations. Their results demonstrated a maximum inhibition efficiency of anti-corrosion coating with 0.5 wt% copper nanoparticle solution, illustrating that their modified coating provided superior corrosion protection and enhanced the coverage of carbon steel.

REFERENCES

1. Lee, J., S. Mahendra, and P.J. Alvarez, Nanomaterials in the construction industry: a review of their applications and environmental health and safety considerations. *ACS Nano*, 2010. **4**(7): pp. 3580–3590.
2. Mohajerani, A., et al., Nanoparticles in construction materials and other applications, and implications of nanoparticle use. *Materials*, 2019. **12**(19): p. 3052.
3. Yang, G. and S.-J. Park, Deformation of single crystals, polycrystalline materials, and thin films: a review. *Materials*, 2019. **12**(12): p. 2003.
4. Teizer, J., et al., Nanotechnology and its impact on construction: bridging the gap between researchers and industry professionals. *Journal of Construction Engineering and Management*, 2012. **138**(5): pp. 594–604.
5. Vance, M.E., et al., Nanotechnology in the real world: redeveloping the nanomaterial consumer products inventory. *Beilstein Journal of Nanotechnology*, 2015. **6**(1): pp. 1769–1780.
6. Murthy, S.K., Nanoparticles in modern medicine: state of the art and future challenges. *International Journal of Nanomedicine*, 2007. **2**(2): p. 129.
7. Rae, A., Real life applications of nanotechnology in electronics. *OnBoard Technology*, 2006. **2006**: p. 28.
8. Raj, S., et al., Nanotechnology in cosmetics: opportunities and challenges. *Journal of Pharmacy & Bioallied Sciences*, 2012. **4**(3): p. 186.
9. Epa, U., Construction and demolition debris: generation in the United States, 2014. Office of Resource Conservation and Recovery, US EPA. 2016.
10. Mohamed, A.S.Y., Nano-innovation in construction, a new era of sustainability. *Energy*, 2015. **65**: p. 28.
11. Chakraborty, S. and G. Bhattacharya, *Proceedings of the International Symposium on Engineering Under Uncertainty: Safety Assessment and Management (ISEUSAM-2012)*, 2013, Springer Science & Business Media.
12. Lee, H.K., et al., Fluctuation of electrical properties of carbon-based nanomaterials/cement composites: case studies and parametric modeling. *Cement and Concrete Composites*, 2019. **102**: pp. 55–70.
13. Du, H., S. Du, and X. Liu, Durability performances of concrete with nano-silica. *Construction and Building Materials*, 2014. **73**: pp. 705–712.
14. Samadi, M., et al., Waste ceramic as low cost and eco-friendly materials in the production of sustainable mortars. *Journal of Cleaner Production*, 2020: p. 121825.
15. Mohammadhosseini, H., et al., Enhanced performance of green mortar comprising high volume of ceramic waste in aggressive environments. *Construction and Building Materials*, 2019. **212**: pp. 607–617.
16. Nazari, A., et al., Influence of Al_2O_3 nanoparticles on the compressive strength and workability of blended concrete. *Journal of American Science*, 2010. **6**(5): pp. 6–9.
17. Jalal, M., Influence of class F fly ash and silica nano-micro powder on water permeability and thermal properties of high performance cementitious composites. *Science and Engineering of Composite Materials*, 2013. **20**(1): pp. 41–46.
18. Shah, K.W. and G.F. Huseien, Biomimetic self-healing cementitious construction materials for smart buildings. *Biomimetics*, 2020. **5**(4): p. 47.

19. Wang, Y., et al., Beneficial effect of nanomaterials on the interfacial transition zone (ITZ) of non-dispersible underwater concrete. *Construction and Building Materials*, 2021. **293**: p. 123472.

20. Singh, N., S. Saxena, and M. Kumar, Effect of nanomaterials on the properties of geopolymer mortars and concrete. *Materials Today: Proceedings*, 2018. **5**(3): pp. 9035–9040.

21. Yang, H., et al., Effects of nano silica on the properties of cement-based materials: a comprehensive review. *Construction and Building Materials*, 2021. **282**: p. 122715.

22. Balapour, M., A. Joshaghani, and F. Althoey, Nano-SiO$_2$ contribution to mechanical, durability, fresh and microstructural characteristics of concrete: a review. *Construction and Building Materials*, 2018. **181**: pp. 27–41.

23. Garcia-Contreras, R., et al., Mechanical, antibacterial and bond strength properties of nano-titanium-enriched glass ionomer cement. *Journal of Applied Oral Science*, 2015. **23**: pp. 321–328.

24. Shafaei, D., et al., Multiscale pore structure analysis of nano titanium dioxide cement mortar composite. *Materials Today Communications*, 2020. **22**: p. 100779.

25. Jin, J., et al., A study on modified bitumen with metal doped nano-TiO$_2$ pillared montmorillonite. *Materials*, 2019. **12**(12): p. 1910.

26. Li, L., et al., Influence of methylcellulose on the impermeability properties of carbon nanotube-based cement pastes at different water-to-cement ratios. *Construction and Building Materials*, 2020. **244**: p. 118403.

27. Shah, S.P., P. Hou, and M.S. Konsta-Gdoutos, Nano-modification of cementitious material: toward a stronger and durable concrete. *Journal of Sustainable Cement-Based Materials*, 2016. **5**(1–2): pp. 1–22.

28. Fang, C., et al., Nanomaterials applied in asphalt modification: a review. *Journal of Materials Science & Technology*, 2013. **29**(7): pp. 589–594.

29. Hamedi, G.H., H. Ghahremani, and D. Saedi, Investigation the effect of short term aging on thermodynamic parameters and thermal cracking of asphalt mixtures modified with nanomaterials. *Road Materials and Pavement Design*, 2020: pp. 1–28.

30. Wu, S. and O. Tahri, State-of-art carbon and graphene family nanomaterials for asphalt modification. *Road Materials and Pavement Design*, 2021. **22**(4): pp. 735–756.

31. Wang, R., et al., Investigating the effectiveness of carbon nanomaterials on asphalt binders from hot storage stability, thermodynamics, and mechanism perspectives. *Journal of Cleaner Production*, 2020. **276**: p. 124180.

32. Hamedi, G.H., Evaluating the effect of asphalt binder modification using nanomaterials on the moisture damage of hot mix asphalt. *Road Materials and Pavement Design*, 2017. **18**(6): pp. 1375–1394.

33. Taherkhani, H., S. Afroozi, and S. Javanmard, Comparative study of the effects of nanosilica and zyco-soil nanomaterials on the properties of asphalt concrete. *Journal of Materials in Civil Engineering*, 2017. **29**(8): p. 04017054.

34. Sarkar, D., M. Pal, and A. Sarkar, Laboratory evaluation of asphalt concrete prepared with over burnt brick aggregate treated by zycosoil. *International Journal of Civil, Environmental, Structural, Construction and Architectural Engineering*, 2014. **8**: pp. 1302–1306.

35. Behbahani, H., et al., Evaluation of performance and moisture sensitivity of glasphalt mixtures modified with nanotechnology zycosoil as an anti-stripping additive. *Construction and Building Materials*, 2015. **78**: pp. 60–68.

36. Mohajerani, A., A.A. Kadir, and L. Larobina, A practical proposal for solving the world's cigarette butt problem: recycling in fired clay bricks. *Waste Management*, 2016. **52**: pp. 228–244.

37. Mohajerani, A., et al., A proposal for recycling the world's unused stockpiles of treated wastewater sludge (biosolids) in fired-clay bricks. *Buildings*, 2019. **9**(1): p. 14.

38. Niroumand, H., M. Zain, and S.N. Alhosseini, The influence of nano-clays on compressive strength of earth bricks as sustainable materials. *Procedia-Social and Behavioral Sciences*, 2013. **89**: pp. 862–865.
39. Stefanidou, M., and A. Karozou, Testing the effectiveness of protective coatings on traditional bricks. *Construction and Building Materials*, 2016. **111**: pp. 482–487.
40. Hegazy, M., A. Badawi, S.A. El Rehim, and W. Kamel, Influence of copper nanoparticles capped by cationic surfactant as modifier for steel anti-corrosion paints. *Egyptian Journal of Petroleum*, 2013. **22**: pp. 549–556.

3 Nanomaterial-Based Cement Concrete

Engineering Properties and Durability Performance

3.1 INTRODUCTION

Globally, concrete is still the most common manufactured material used in terms of volume. Except for water, concrete is consumed in the largest amount by mass in the world [1]. According to industry statistics published in *The Global Cement Report, 13th Edition* [2], the cement consumption increased globally by 2.8% from 4.08 (Mt) in 2016 to 4.2 (Mt) in 2019. The global demand for cement is set to grow by 12%–23% by 2050 compared to 2014 [3] as a result of foreseen growth in countries such as China and India and regions such as Southeast Asia, Middle East, and Northern Africa. As economies expand in emerging and developing nations, there is a growing need to construct new infrastructures, such as bridges, highways, schools, hospitals, canals, tunnels, pavement, and sidewalks, all of which demands an intensive use cement and concrete [4,5]. Ordinary Portland cement (OPC) is made by heating limestone or chalk with clay in a rotary kiln at high temperatures (1450°C–1650°C), resulting in solid clinker. This is then grinded with a small amount (3%–5%) of gypsum to produce conventional cement. Besides consuming high energy because of heating process, rapid landscape degradation, dust production through transportation process, noise generation in quarries, and raw material production are all major contributors of environmental issues associated with OPC manufacturing [6]. In recent years, carbon dioxide (CO_2) emission created a new challenge for the cement industry. Approximately 1 tonne of cement produces around 1 tonne of CO_2. Cement production is currently responsible for about 7.1% of global CO_2 emissions and 2%–3% of energy usage. Emissions from cement production will increase in levels of 13.5% (2.596 Mt) and 5.7% (2.416 Mt) above the 2010 levels by 2030 and 2050, respectively [7]. While cement produces massive quantities of greenhouse gases (CO_2, SOx, and NOx) emissions [8,9], dwindling amounts of limestone, and a large amount of energy consumption, several negative effects can be observed to be associated with the manufacturing process. These release a large amount of dust into the atmosphere which has a detrimental impact on the respiratory system of living beings.

Due to the advancement in instrumentation technology for observation and measurement at the nanoscale, the concept of nanotechnology has become prevalent in various fields of science. This refers to rebuilding the matter structures (at the nanoscale) while keeping their fundamental properties and functions [10].

DOI: 10.1201/9781003196143-3

Two techniques have been introduced. The first approach is from top to bottom [11,12] while the second is from bottom to top [13]. The selection of the two methods is based on several factors such as suitability, cost, and expertise related to nano-behaviour [14]. One of the techniques (top to down) uses milling to reduce sizes to nanoscale while retaining their original properties or chemical structure without any modification on atomic level. On the other hand, the down to top approach is focused on the design and control of materials' atoms using chemical synthesis technique [13–15]. While the top to down approach is considered more effective in cost and complexity, the down to top approach provides a more standardized and neat arrangement of nanomaterials, but it is costly and complicated operation [10]. By employing the nanotechnology potential and using industrial by-products such as coal fly ash, iron blast-furnace slag, and nanomaterials, both CO_2 emissions and energy can be minimized or reduced, which could help to achieve the sustainability requirements (environment, society, and cost) in the construction industry.

3.2 NANOMATERIAL-MODIFIED CEMENT BINDER

Using nanoparticles in concrete plays a key role to enhance its microstructure and bulk properties. Since nanoparticles have a large surface area to volume ratio, their presence encourages the pozzolanic chemical reaction. This reaction increases the temperature which accelerates the hydration process and reduced the setting time of the matrix. Morphology can be enhanced by improving the interfacial contacts between the aggregates and the binder. Nanoparticles act as fillers which can result in the reduction of porosity and improving packing model structure. The enhanced morphology reflects positively on the bulk properties and durability of concrete.

Most of the research studies on the nano-modification of cement and concrete incorporated nano-silica (nano-SiO_2) [15–18]. Results showed improvement in cement hydration reaction rate and microstructure leading to high bulk density, early strength, and better resistance to chemical attacks. Besides that, it helps to limit the leaching of calcium hydroxide ($Ca(OH)_2$) which is considered responsible for a significant part of concrete degradation, thereby increasing the concrete's service life. SiO_2 nanoparticles act not only as a filler to improve the microstructure and interfacial interaction but also contribute to pozzolanic reactions, which carry a significant effect on the cement concretes' performance. Nanomaterials, such as nano-alumina (nano-Al_2O_3) [19–25], nano titanium dioxide (nano-TiO_2) [26–30], carbon nanotubes (CNTs) [31,32], and nano clay [33], have also been investigated. The results reflected a remarkable increase both in short-term strength and long-term strength and better performance in harmful environments. The reasons for these enhancements are that nanoparticles improve the microstructure and reinforce the bond between the cement hydration products and the aggregates [34].

Nano-alumina (nano-Al_2O_3) showed comparable results to those of nano-silica (nano-SiO_2) particles on the concrete, including improved effects in strength and durability. Even new functional advantages can be found by nano-TiO_2, such as photo catalysis and piezoelectricity by using CNTs [35–37]. It was observed that CNTs/

nanofibers (namely CNTs and CNFs) would turn into the most valuable nanomaterials for further improving the mechanical properties in cement-based materials while also providing novel properties such as self-sensing capabilities and electromagnetic shielding capabilities [32,38].

3.3 FRESH PROPERTIES

Remarkable effects of nanomaterials on fresh properties of concrete were recorded. One of them is a significant increase in heat of hydration mixture modified with incorporating nanoparticles. This increase in the hydration peak temperature is a strong indicator of the role of nanomaterials in accelerating the reaction rate which leads to minimize the setting time [39,40]. According to Land and Stephan [41,42] the increased surface area is mainly responsible for the increase in hydration heat. Several research studies reported that the heat of hydration increased as the nanoparticles' dosage increased which means more reduction in setting time of the matrix [25,39,43]. Qing et al. [44] reported that the higher number of finer particles and high specific surface area of nanoparticles have a significant effect on workability and setting times which makes the mixture more cohesive and requires more water content for wetting its surface. The authors remarked that increasing the specific surface area of nanomaterials products would increase the wettable surface area of cement paste and thus more amount of water is required in adsorption process. Mukharjee and Barai [45] found that the setting time is strongly related to the degree of fineness of cement materials. The higher amount of fineness of nanocementing materials improved the cement's reactivity. In addition to that, nanoparticles serve as nucleation sites and thus accelerate the hydration rate, which leads to shorter setting times. Phoo-ngernkham et al. [25,46] stated that with an increase in nano-SiO_2, pastes set faster with readily available free calcium ions released from cement or slag to form additional C-S-H. From the previous studies, it can be noted that the workability of plain pastes/mortars/concretes decreased with increasing nanomaterial contents. The addition of (SiO_2, SF, Al_2O_3, TiO_2, and NC) nanoparticles reduced remarkably the flowability [47,48] due to their higher fineness that leads to gain the cohesiveness in the paste. Hosseini et al. [49] recorded a remarkable reduction in workability by adding nanoparticles. The slump flow reduced to 47.1% and 70.59% by adding 1.5% and 3% colloidal NS to 100% recycled coarse aggregate-modified concrete. They also declared the reduction in workability to the high surface area of the nano-silica particles. Jalal et al. [50] and Collepardi et al. [51] mentioned that the presence of silica fume (SF) and nano-silica (NS) decreased the flowability; however, they would improve several other fresh concrete's characteristics such as good consistency and less bleeding and segregation in the mixtures. Joshaghani et al. [52] found that all the TiO_2, Al_2O_3, and Fe_2O_3 nanoparticles caused remarked reduction in slump flow. It is worth to note in this respect that with the increase in nanomaterials' percentage, the percentage of high-range water reducing admixture (HRWRA) increased to achieve the slump flow of 650 ± 25 mm until it reached 36% in the case of 5% TiO_2. Most researchers attributed the reduction in flowability to higher specific surface area of nanoparticles than that of cement, because it absorbs more water which reduces the slump of concrete [53].

3.4 STRENGTH PERFORMANCE

The use of nanomaterials in cementitious systems has drawn attention for the past two decades due to their effects in improving mechanical strength. A group of researchers have investigated the effect of deferent nanomaterials on the strength properties of a nano-modified concrete system [31,54–62]. Most of their results reflected improvement in compressive, flexural, and split strength, especially in the early ages, by adding or replacing cement in a small dosage of nano-Al_2O_3, Fe_2O_3, Fe_3O_4, ZnO_2, ZrO_2, Cu_2O_3, CuO, $CaCO_3$, SF, CTs, and clay.

Zhang et al. [63] studied separately the replacement effects of 1% of nano-SiO_2 (15 and 50 nm), nano-Fe_2O_3 (50 nm), and nano-NiO (15 nm) on hydration and strength properties of white cement (W/C = 0.5). The results at 3 days of curing age revealed that the compressive strength of samples (15 nm SiO_2), (50 nm SiO_2), (15 nm NiO), and (50 nm Fe_2O_3) increased 21.9%, 20.5%, 22.8%, and 23.9%, respectively, compared to that of control sample. At 28 days, the compressive strength increase rate was 10.2%, 9.5%, 10.5%, and 11.2%, respectively. The researchers attributed these increases to the pozzolanic reaction activity, filling effect and nucleation effect of NS, low calcium hydroxide content, and reduced porosity which reflected positively on the compressive strength.

Said et al. [40] investigated the effect of nano-silica on both pure cement concrete and 30% fly ash cement replacement. The finding indicated that adding 3% and 6% of nano-silica improved the compressive strength after 3, 7, and 28 days of curing by 18%, 14%, and 36% respectively, for both specimens. The authors ascribed the strength properties' improvements to pozzolanic reaction and filler effects of nano-silica.

Kaur et al. [64] studied the influence of nano metakaolin (NMK) on compressive strength and microstructure of fly ash-based mortar by replacing 0%, 2%, 4%, 6%, 8%, and 10% of fly ash by NMK. After 3 days of curing, results revealed that 4% replacement sample gained about 70%–80% of its 28 days' compressive strength. This increment in compressive strength compared to control sample was 26.5%, 21.4%, 21.4%, and 22.7% for 3, 7, 14, and 28 days respectively. Increasing the NMK content to 10%, a reduction in compressive strength by around 1%–2% was observed compared to control sample for all curing ages. These increments in compressive strength could be due to the contribution of silica and alumina oxide in NMK in the pozzolanic reaction.

Gunasekara et al. [65] studied the effect of nano-silica addition in high-volume fly ash (HVFA) hydrated lime blended concrete. The results showed that by incorporating 3% nano-silica into HVFA (65% and 80%) mixture increased compressive strength by 50% and 98.6%, respectively, at 7 days, and 10.3% and 35.9% strength gain at 28 days, compared to the control samples. This is attributed to very fine nano-silica particles which served as nucleation sites for the hydration of cement particles. Also, a reduction in calcium hydroxide (CH) was observed by adding NS to concrete indicating the additional formation of C–S–H in the system at 7 and 28 days, which resulted in the increase in compressive strength.

Zhen et al. [66] studied the effectiveness of nano-titanium dioxide on mechanical strength and microstructure properties of reactive powder concrete. The authors recorded that using 2% nano-TiO_2 enhanced the compressive strength by 18.5% at

28 days. The results also showed enhancement in flexural strength by 52.72% and 47.07% at 3 and 28 days, respectively. The improvements in compressive and flexural strengths of concrete could be due to the fact that it can increase its toughness. Two factors may have worked together for the enhancing effect. On the one hand, the hydration products of cement deposit on the nanoparticles due to their huge surface activity and growth to form conglomerations containing the nanoparticles as 'nucleus'. This effect is called the nucleation effect.

Liu et al. [67] studied the effects of nano-SiO_2 on early age's strength of steam-cured HVFA cement system. The results showed that the addition of 1% nano-SiO_2 improved the compressive and flexural strengths by 22% and 13%, respectively, in comparison with reference. Also, addition of 4.0% NS led to 106% and 67% increase, respectively. The authors described the improvement in compressive and flexural strengths to the pozzolanic reaction of nano-SiO_2 with CH forming additional C-S-H gel. Once the NS dissolves in water, it forms H_2SiO_4 which reacts with Ca^{2+} to form C-S-H gel. Also, NS acts as the seed in pores providing more nucleation.

Fallah et al. [68] reported that replacement of cement by 1%, 2%, and 3% of nano-SiO_2 increased tensile strength by 12.96%, 7.82%, and 16.10%, respectively, compared to the control sample containing only Portland cement. This increase in tensile strength can be attributed to the improvement of bonds between the hydration cement products and the aggregates and could be due to the conversion of CH to C-S-H gel in the presence of nano-silica.

Lim et al. [62] investigated the influence of SiO_2-Al_2O_3 on strength and microstructure by using high percentages (20%, 40%, 60%) of nanoparticles of ceramic waste in mortar as a cement replacement. At ages of 28 and 90 days, the specimens containing 20% and 40% of ceramic powder as a cement replacement showed higher compressive strength compared to the control sample. The author described the improvement in compressive strength to the pozzolanic reaction between nano-SiO_2 particles and calcium hydroxide $(Ca(OH)_2)$ released at the earlier stage of cement hydration process. On the other hand, the mortar containing 60% of Al_2O_3–SiO_2 nanoparticles showed lower compressive strength compared to that of the mortar with 20% and 40% replacement. The decrease in compressive strength of 60% mixture could be attributed to the reduction in the cement binder content and lower workability.

On [69], authors investigated the influence of glass bottles waste nano powder (BGWNP) on alkali-activated mortars (AAMs). The author recorded that the replacement of ground blast furnace slag by 0%–5% nano glass powder improved the split strength by 22.22%, and by increasing it to 10% wt, improved the split strength by 33.33% compared to the control sample at 28 days of curing. The same study recorded the increase in flexural strengths by 14% with 5%, respectively. The author attributed the increase in flexural and split strengths of nano-powder's role in improving the microstructure of mortar.

3.5 MICROSTRUCTURE PROPERTIES

In this section, detailed effects of nanomaterials on the mechanical properties of concrete are discussed from microstructure perspective. Most researchers reported that the increase in compressive strength with very small dosages of nanoparticles owed to the role of these finer particles. These particles provide further nucleation sites for pozzolanic reaction which generate additional C-S-H gel, thus leading to an increase in compressive strength [62]. Other researchers attributed the increase in compressive strength due to improvement in interfacial transaction zones (ITZs). In other words, the addition of nanoparticles prevents large pores to create and consume the $Ca(OH)_2$ formed in ITZ. Furthermore, unreacted NS particles act as fillers, making the microstructure denser with more homogenous and better interlock morphology [53,70]. The extremely fine nanoparticles provide more nucleation sites which create a denser and more compacted matrix that reflects positively on the compressive strength. The growth rate of $Ca(OH)_2$ crystals is reduced by the addition of nanomaterials due to high specific surface area and high reactivity. In this way, the materials can reduce the amount of $Ca(OH)_2$ crystals that are formed in the ITZ and generate more C-S-H gel to fill the voids [53]. The improvement in ITZ with the addition of nanoparticles is a clear evidence on the consumption of $Ca(OH)_2$ by nanoparticles [71]. Making the matrix more compact and improved ITZ between the binder and aggregates also reflects positively on the compressive strength performance. The increase in compressive strength of nanomaterial-modified concrete is increased with the dosage per cent of nanomaterial content towards the optimum content. Above the optimum percentage, a higher amount of nanomaterial leads to a decrease in compressive strength [17,69]. It was found the increasing dosage percentage of nanoparticles in concrete and once it exceeds the optimum causes more agglomeration sites which produce week zones and voids, resulting in a decrease in strength. According to Elkady et al. [46], the increment or reduction in compressive strength depends on the agglomeration degree of nanoparticles. In other words, if the per cent content of nanoparticles exceeds the optimum, the particles will physically stick to each other creating weak points in the matrix [72]. The same conclusions were reported by Li et al. [27] who reported that using a higher percentage of nanoparticles in concrete caused irregularities in the dispersion of particles (10–5 nm size) which resulted in weak zones. Other studies reported that the reason for reduced compression strength might be due to the reduction in crystalline content of $Ca(OH)_2$ which is considered the fundamental compound to react with the nanoparticles to form additional C-S-H gel [62].

According to previously reviewed studies which relate to the effect of nanoparticles on the development or reduction of compressive strength may also apply to tensile and split strength. In other studies, it was reported that by adding nanoparticles, the flexural and tensile strengths enhance significantly [69]. This is because of pores' filling behaviour and the pozzolanic reaction of nanoparticles [72]. The improvements in early splitting and tensile strengths by incorporating nanomaterials in nano-modified concrete are due to their high reactivity which accelerates the consumption of calcium hydroxide produced during the hydration of Portland cement and refines the progressed micro-cracks at the early ages. As a result, cement hydration is accelerated by adding a small dosage of nanoparticles forming reaction products which promote better density and reduce pore size in the cement matrix [50].

In short, results indicate that the split and tensile strengths of nano-modified concrete increase significantly by adding nanoparticles. The effects of nanoparticles in nano-modified concrete could be listed in the following four points:

1. Nanomaterials act as pore fillers which allow to increase matrix compaction. The kinetics and hydration of cement are expected to influence substantially by the addition of a small amount of nanoparticles. These allow better voids' filling due to their high specific area and increased electrostatic force.
2. Nanoparticles with high specific surface accelerate the hydration process of cement matrix, which leads to generating C-S-H gel clusters of calcium silica.
3. Nanoparticles increase the homogeneity of cement matrix and make it more compacted compared to the conventional samples.
4. As a result of nanoparticles' ability to react chemically with $Ca(OH)_2$, their high reactivity accelerates the consumption and accumulates in the micropores and ITZ which induce refining the progressed micro-cracks at early ages and improving matrix's microstructure.

3.6 DURABILITY PROPERTIES

As expected, nanomaterials can inevitably improve the concrete's durability because of their role in improving the concrete microstructure. Analysing the durability's parameters, such as porosity, chloride ion penetration, and chemical attacks of the modified concrete containing nanomaterials, several studies observed that they have a positive effect on concrete's durability. Porosity represents the volume and connectivity of pores. Indirectly, lower porosity indicates a lower risk probability of chemical attacks due to chloride and sulphate ions. Considering the porosity and defected ITZ which are most critical factors for any deformation or defect in concrete's durability, several research studies have focused on nanomaterials' role to overcome this challenge. These studies reported that by adding small doses of nanomaterials with uniform distribution of mixture, the porosity of concrete improves significantly [73,74]. Also, they can provide additional nucleation sites which contribute to a decrease in the pores and enhance the compaction by filling them as well as the gaps in C-S-H gel of the matrix. These factors lead to reduce the porosity and generate better performance and durability [52]. It was found that addition of small amount of nanomaterials is more effective in improving the porosity by refining the micropore spaces which reflect positively on the durability properties. Besides that, these fine particles of nanomaterials located as a kernel lead to accelerate the hydration reaction, they also behave as a main reactant due to their high reactivity, creating more C-S-H or C-S-A-H gel. This makes the matrix more homogeneous and more compacted. The mechanism by which the nanomaterials improve the pore structure of concrete can be illustrated [50] by supposing that they are uniformly dispersed throughout the concrete and are contained in a cube pattern, then the distance between them could be said is equal. Once the hydration process begins, the hydration products start working to wrap these particles and distribute uniformly into the matrix [75]. If the nanoparticle contents and the distance between them are adequate, the crystallization is controlled to be appropriate by limiting the growth of calcium hydroxide. Furthermore, due to their reactivity and high surface area, the nanoparticles present in the cement paste as the kernel could promote the hydration process of cement, thus improving the microstructure of concrete matrix better especially, at early ages. Other researchers lauded nanoparticles for their role in improving the ITZ in cement matrix samples due to their role to convert the $Ca(OH)_2$ crystals to additional C-S-H gel.

Likewise, the carbonation issue is considered to be one of the most critical detrimental deterioration phenomena due to its extremely deleterious impact on hardened properties of concrete. It's worth noting here that the carbonation and the presence of chloride ions are the primary causes of steel corrosion in concrete structures. Behfarnia et al. [76] reported that as a result of carbonation, the alkalinity of concrete decreases from pH 12.6 to 11; the protective cover on the reinforcement steel bar is cracked, which allows oxidizing agents to reach to the steel bars where the oxidation and reduction process takes place. The occurrence of carbonization phenomenon in concrete and the role of nanoparticles to reduce its negative effects can be explained as follows: in the presence of moisture, atmospheric CO_2 penetrates into the concrete through small pores and reacts with alkaline hydrated products forming $CaCO_3$. In the case of acidic rain, H_2CO_3 acid reacts with alkaline hydrated cement products and also forms $CaCO_3$ [77]. In the presence of nanomaterials, they work on refining cracks in concrete and increase its compactness "as discussed previously". This makes it difficult for H_2CO_3 acid and CO_2 to penetrate into the concrete through micro-cracks, thus reducing the possibility of carbonation phenomenon.

Group of researchers [77–79] studied the effects of nanomaterial additives on the carbonation phenomenon of concrete. Most of their results showed that a small dosage of nanomaterials may play a notable positive role in concrete's resistance to the carbonation phenomenon. It was observed that the specimens containing a small dosage of nanoparticles exhibited a lower carbonation depth values than the conventional concrete. Thus, their addition enhances significantly the carbonation resistance which could be attributed to the improvement in the microstructure of concrete/mortar matrix [80–82].

The protective effects of nanoparticles against chloride and sulphate ions attacks were recorded as remarkable. Several studies mentioned that by adding small dosages of nanomaterials to concrete systems, the chloride ion penetration decreased drastically [52,73,74,83–88]. Said et al. [40] recorded that adding 6% nano-silica reduced the chloride penetration depth by 59.2%–69.6%. The average penetration depth was 10.3, 3.1, and 4.6 mm for 0%, 3%, and 6% nano-silica addition to pure cement concrete while these values were 8.1, 3.1, and 3.3 mm, respectively, when the cement was replaced by 30% fly ash.

According to the findings of previous studies, nanomaterials reflected a significant effect on the reduction of sulphate and chloride ion penetration of modified concrete. This reduction could be attributed to the high reactivity of nanoparticles that promotes their reaction between them and calcium hydroxide which lowers the resistance against chemical attacks. In other words, the strong resistance against chemical attacks of nano-modified concrete is due to their filling the gaps in C–S–H gel and behaves as a nucleation, so, the micropores and cracks could be refined and ITZ improved. Thus, the resistance against sulphate and chloride ions penetration will be improved [89,90].

Samadi et al. [91] evaluated the mechanical performance of concrete/mortars containing nanomaterials after exposure to chemical attacks. It was reported that the samples containing nano-palm oil maintained their compressive strength by 18.6% higher than the control sample after 18 months of immersion in 5% sodium sulphate solution. Furthermore, other than the control specimen, the nano-palm oil fuel ash (NPOA)-modified specimen did not show any change in size and maintained its microstructure without showing any cracks. The authors attributed the improvement against sulphate resistance attack to the formation of low gypsum, ettringite, and thaumasite and filling the microspores.

It can be concluded that the effectiveness of nanomaterials in reducing porosity, repairing micro-cracks, and improving ITZ played a significant role in increasing concrete's resistance to chemical ion attack and thus enhancing the durability. Similar negative effects on concretes' durability exist when the percentage of nanomaterials is above the optimum level.

3.7 SUMMARY

The mechanisms' effects of nanomaterials on the setting time, slump, mechanical performance, and durability of concrete are described in detail in this chapter. After reviewing the exhaustive research studies on nanomaterials' concrete, the following conclusions can be drawn:

- Incorporation of nanoparticles into the mix resulted in a reduction in setting time and workability. This reduction is attributed to the high surface area and reactivity of the nanoparticles.
- The nanoparticles played an important role in improving the strength performance of concrete, especially at early ages due to their reaction with $Ca(OH)_2$ to produce more C-S-H gel, and filling the matrix pores.
- Addition of nanoparticles in concrete enhanced the durability by increasing their resistance against carbonization and chemical attacks due to their improved microstructure, filling the micro-cracks and reduction in porosity.
- Due to their high surface area and high reactivity, the incorporation of nanomaterials accelerated the pozzolanic reaction process which led to increase significantly the mechanical properties.
- The nanomaterials react with calcium hydroxide to produce additional C-S-H gel. Besides that, the nanoparticles acted as fillers or nucleation sites which improved the microstructure of the modified concrete.
- Increased use of OPC resulted in a serious environmental damage. In terms of sustainability, energy savings, and environmental issues, huge advantages and usefulness of nano-concrete were demonstrated. The current and

future trends to substitute conventional concrete with nano-concrete are for the future sustainability.

- Increasing the percentage of nanomaterials' dosage by more than the optimum reflected negatively on the strength and durability properties of concrete. These are due to their difficulty to disperse uniformly and the formation of weak zones within the matrix.

REFERENCES

1. Monteiro, P., S. Miller, and A. Horvath, Towards sustainable concrete. *Nature Materials*, 2017. **16**: pp. 698–699.
2. Ltd., T.P. The global cement report™ – 13th edition. UK and the EU. 2019; Available from: https://www.cemnet.com/Publications/Item/182291/the-global-cement-report-13th-edition.html.
3. IEA. Cement technology roadmap plots path to cutting CO_2 emissions 24% by 2050. 6 April 2018; Available from: https://www.iea.org/news/cement-technology-roadmap-plots-path-to-cutting-co2-emissions-24-by-2050.
4. Cao, Z., et al., Elaborating the history of our cementing societies: an in-use stock perspective. *Environmental Science & Technology*, 2017. **51**(19): pp. 11468–11475.
5. Monteiro, P.J.M., S.A. Miller, and A. Horvath, Towards sustainable concrete. *Nature Materials*, 2017. **16**(7): pp. 698–699.
6. Huseien, G.F., et al., Geopolymer mortars as sustainable repair material: a comprehensive review. *Renewable and Sustainable Energy Reviews*, 2017. **80**: pp. 54–74.
7. Belbute, J.M. and A.M. Pereira, Reference forecasts for CO_2 emissions from fossil-fuel combustion and cement production in Portugal. *Energy Policy*, 2020. **144**: p. 111642.
8. Pacyna, E.G., et al., Global emission of mercury to the atmosphere from anthropogenic sources in 2005 and projections to 2020. *Atmospheric Environment*, 2010. **44**(20): pp. 2487–2499.
9. Streets, D., et al., Anthropogenic mercury emissions in China. *Atmospheric Environment*, 2005. **39**(40): pp. 7789–7806.
10. Sanchez, F. and K. Sobolev, Nanotechnology in concrete – a review. *Construction and Building Materials*, 2010. **24**(11): pp. 2060–2071.
11. Abdoli, H., et al., Effect of high energy ball milling on compressibility of nanostructured composite powder. *Powder Metallurgy*, 2011. **54**(1): pp. 24–29.
12. Huseien, G.F., K.W. Shah, and A.R.M. Sam, Sustainability of nanomaterials based self-healing concrete: an all-inclusive insight. *Journal of Building Engineering*, 2019. **23**: pp. 155–171.
13. Jankowska, E. and W. Zatorski. Emission of nanosize particles in the process of nanoclay blending. in 2009 *Third International Conference on Quantum, Nano and Micro Technologies*, 2009, IEEE.
14. Sanchez, F. and K. Sobolev, Nanotechnology in concrete–a review. *Construction and Building Materials*, 2010. **24**(11): pp. 2060–2071.
15. Li, G., Properties of high-volume fly ash concrete incorporating nano-SiO_2. *Cement and Concrete Research*, 2004. **34**(6): pp. 1043–1049.
16. Ghafari, E., et al., The effect of nanosilica addition on flowability, strength and transport properties of ultra high performance concrete. *Materials & Design*, 2014. **59**: pp. 1–9.
17. Seifan, M., S. Mendoza, and A. Berenjian, Mechanical properties and durability performance of fly ash based mortar containing nano- and micro-silica additives. *Construction and Building Materials*, 2020. **252**: p. 119121.

18. Khaloo, A., M.H. Mobini, and P. Hosseini, Influence of different types of nano-SiO_2 particles on properties of high-performance concrete. *Construction and Building Materials*, 2016. **113**: pp. 188–201.

19. Adak, D., M. Sarkar, and S. Mandal, Effect of nano-silica on strength and durability of fly ash based geopolymer mortar. *Construction and Building Materials*, 2014. **70**: pp. 453–459.

20. Li, Z., et al., Investigations on the preparation and mechanical properties of the nano-alumina reinforced cement composite. *Materials Letters*, 2006. **60**(3): pp. 356–359.

21. Nazari, A., et al., Influence of Al_2O_3 nanoparticles on the compressive strength and workability of blended concrete. *Journal of American Science*, 2010. **6**(5): pp. 6–9.

22. Nazari, A., et al., Mechanical properties of cement mortar with Al_2O_3 nanoparticles. *Journal of American Science*, 2010. **6**(4): pp. 94–97.

23. Hase, B. and V. Rathi, Properties of high strength concrete incorporating colloidal nano-Al_2O_3. *International Journal of Innovative Science Engineering and Technology*, 2015. **4**(3): pp. 959–963.

24. Behfarnia, K. and N. Salemi, The effects of nano-silica and nano-alumina on frost resistance of normal concrete. *Construction and Building Materials*, 2013. **48**: pp. 580–584.

25. Phoo-ngernkham, T., et al., The effect of adding nano-SiO_2 and nano-Al_2O_3 on properties of high calcium fly ash geopolymer cured at ambient temperature. *Materials & Design*, 2014. **55**: pp. 58–65.

26. Massa, M.A., et al., Synthesis of new antibacterial composite coating for titanium based on highly ordered nanoporous silica and silver nanoparticles. *Materials Science and Engineering: C*, 2014. **45**: pp. 146–153.

27. Li, H., M.-h. Zhang, and J.-p. Ou, Abrasion resistance of concrete containing nano-particles for pavement. *Wear*, 2006. **260**(11–12): pp. 1262–1266.

28. Li, H., M.-h. Zhang, and J.-p. Ou, Flexural fatigue performance of concrete containing nano-particles for pavement. *International Journal of Fatigue*, 2007. **29**(7): pp. 1292–1301.

29. Sorathiya, J., S. Shah, and S. Kacha, Effect on addition of nano "titanium dioxide" (TiO_2) on compressive strength of cementitious concrete. 2018. **1**: pp. 219–211.

30. Jayapalan, A., B. Lee, and K. Kurtis, Effect of nano-sized titanium dioxide on early age hydration of Portland cement, in Bittnar Z., Bartos P.J.M., Němeček J., Šmilauer V., Zeman J. (eds) *Nanotechnology in Construction 3*. 2009, Springer: Berlin, Heidelberg, pp. 267–273.

31. Morsy, M., S. Alsayed, and M. Aqel, Hybrid effect of carbon nanotube and nano-clay on physico-mechanical properties of cement mortar. *Construction and Building Materials*, 2011. **25**(1): pp. 145–149.

32. Stynoski, P., P. Mondal, and C. Marsh, Effects of silica additives on fracture properties of carbon nanotube and carbon fiber reinforced Portland cement mortar. *Cement and Concrete Composites*, 2015. **55**: pp. 232–240.

33. Mohamed, A.M., Influence of nano materials on flexural behavior and compressive strength of concrete. *HBRC Journal*, 2016. **12**(2): pp. 212–225.

34. Beigi, M.H., et al., An experimental survey on combined effects of fibers and nanosilica on the mechanical, rheological, and durability properties of self-compacting concrete. *Materials & Design*, 2013. **50**: pp. 1019–1029.

35. Liu, Z.G., et al., Piezoresistive properties of cement mortar with carbon nanotube. *Advanced Materials Research*, 2011. **284**: pp. 310–313, Trans Tech Publications Ltd.

36. Konsta-Gdoutos, M.S., Z.S. Metaxa, and S.P. Shah, Highly dispersed carbon nanotube reinforced cement based materials. *Cement and Concrete Research*, 2010. **40**(7): pp. 1052–1059.

37. Folli, A., et al., TiO_2 photocatalysis in cementitious systems: insights into self-cleaning and depollution chemistry. *Cement and Concrete Research*, 2012. **42**(3): pp. 539–548.

38. Mudimela, P.R., et al., Synthesis of carbon nanotubes and nanofibers on silica and cement matrix materials. *Journal of Nanomaterials*, 2009. **2009**: pp. 1–5.
39. Zhang, M.-H. and J. Islam, Use of nano-silica to reduce setting time and increase early strength of concretes with high volumes of fly ash or slag. *Construction and Building Materials*, 2012. **29**: pp. 573–580.
40. Said, A.M., et al., Properties of concrete incorporating nano-silica. *Construction and Building Materials*, 2012. **36**: pp. 838–844.
41. Land, G. and D. Stephan, The influence of nano-silica on the hydration of ordinary Portland cement. *Journal of Materials Science*, 2012. **47**(2): pp. 1011–1017.
42. Land, G. and D. Stephan, Controlling cement hydration with nanoparticles. *Cement and Concrete Composites*, 2015. **57**: pp. 64–67.
43. Gao, K., et al., Effects of nano-SiO_2 on setting time and compressive strength of alkaliactivated metakaolin-based geopolymer. *The Open Civil Engineering Journal*, 2013. **7**(1): pp. 84–92.
44. Qing, Y., et al., Influence of nano-SiO_2 addition on properties of hardened cement paste as compared with silica fume. *Construction and Building Materials*, 2007. **21**(3): pp. 539–545.
45. Mukharjee, B.B. and S.V. Barai, Assessment of the influence of nano-silica on the behavior of mortar using factorial design of experiments. *Construction and Building Materials*, 2014. **68**: pp. 416–425.
46. Chindaprasirt, P., et al., Effect of SiO_2 and Al_2O_3 on the setting and hardening of high calcium fly ash-based geopolymer systems. *Journal of Materials Science*, 2012. **47**(12): pp. 4876–4883.
47. Gao, X., Q.L. Yu, and H.J.H. Brouwers, Characterization of alkali activated slag–fly ash blends containing nano-silica. *Construction and Building Materials*, 2015. **98**: pp. 397–406.
48. Nazari, A. and S. Riahi, RETRACTED: Al_2O_3 nanoparticles in concrete and different curing media. *Energy and Buildings*, 2011. **43**(6): pp. 1480–1488.
49. Hosseini, P., A. Booshehrian, and A. Madari, Developing concrete recycling strategies by utilization of nano-SiO_2 particles. *Waste and Biomass Valorization*, 2011. **2**(3): pp. 347–355.
50. Jalal, M., et al., Comparative study on effects of Class F fly ash, nano silica and silica fume on properties of high performance self compacting concrete. *Construction and Building Materials*, 2015. **94**: pp. 90–104.
51. Collepardi, M., et al., Influence of amorphous colloidal silica on the properties of selfcompacting concretes. in *Proceedings of the International Conference* "Challenges in Concrete Construction-Innovations and Developments in Concrete Materials and Construction", Dundee, Scotland, 2002.
52. Joshaghani, A., et al., Effects of nano-TiO_2, nano-Al_2O_3, and nano-Fe_2O_3 on rheology, mechanical and durability properties of self-consolidating concrete (SCC): an experimental study. *Construction and Building Materials*, 2020. **245**: p. 118444.
53. Sumesh, M., et al., Incorporation of nano-materials in cement composite and geopolymer based paste and mortar – a review. *Construction and Building Materials*, 2017. **148**: pp. 62–84.
54. Vikulin, V.V., M.K. Alekseev, and I.L. Shkarupa, Study of the effect of some commercially available nanopowders on the strength of concrete based on alumina cement. *Refractories and Industrial Ceramics*, 2011. **52**(4): pp. 288–290.
55. Rashad, A.M., A synopsis about the effect of nano-Al_2O_3, nano-Fe_2O_3, nano-Fe_3O_4 and nano-clay on some properties of cementitious materials – a short guide for Civil Engineer. *Materials & Design*, 2013. **52**: pp. 143–157.

56. Rashad, A.M., Effects of ZnO_2, ZrO_2, Cu_2O_3, CuO, $CaCO_3$, SF, FA, cement and geothermal silica waste nanoparticles on properties of cementitious materials–a short guide for Civil Engineer. *Construction and Building Materials*, 2013. **48**: pp. 1120–1133.

57. Zhang, R., et al., Influences of nano-TiO_2 on the properties of cement-based materials: hydration and drying shrinkage. *Construction and Building Materials*, 2015. **81**: pp. 35–41.

58. Wang, L., H. Zhang, and Y. Gao, Effect of TiO_2 nanoparticles on physical and mechanical properties of cement at low temperatures. *Advances in Materials Science and Engineering*, 2018. **2018**: pp. 1–12.

59. Wang, L., et al., Effect of nano-SiO_2 on the hydration and microstructure of Portland cement. *Nanomaterials (Basel)*, 2016. **6**(12): pp. 1–15.

60. Khotbehsara, M.M., et al., Effect of nano-CuO and fly ash on the properties of self-compacting mortar. *Construction and Building Materials*, 2015. **94**: pp. 758–766.

61. Nazari, A. and S. Riahi, The effects of SiO_2 nanoparticles on physical and mechanical properties of high strength compacting concrete. *Composites Part B: Engineering*, 2011. **42**(3): pp. 570–578.

62. Lim, N.H.A.S., et al., Microstructure and strength properties of mortar containing waste ceramic nanoparticles. *Arabian Journal for Science and Engineering*, 2018. **43**(-10): pp. 5305–5313.

63. Zhang, A., et al., Comparative study on the effects of nano-SiO_2, nano-Fe_2O_3 and nano-NiO on hydration and microscopic properties of white cement. *Construction and Building Materials*, 2019. **228**: p. 116767.

64. Kaur, M., J. Singh, and M. Kaur, Microstructure and strength development of fly ash-based geopolymer mortar: role of nano-metakaolin. *Construction and Building Materials*, 2018. **190**: pp. 672–679.

65. Gunasekara, C., et al., Effect of nano-silica addition into high volume fly ash–hydrated lime blended concrete. *Construction and Building Materials*, 2020. **253**: p. 119205.

66. Li, Z., et al., Effect of nano-titanium dioxide on mechanical and electrical properties and microstructure of reactive powder concrete. *Materials Research Express*, 2017. **4**(9): p. 095008.

67. Liu, M., H. Tan, and X. He, Effects of nano-SiO_2 on early strength and microstructure of steam-cured high volume fly ash cement system. *Construction and Building Materials*, 2019. **194**: pp. 350–359.

68. Fallah, S. and M. Nematzadeh, Mechanical properties and durability of high-strength concrete containing macro-polymeric and polypropylene fibers with nano-silica and silica fume. *Construction and Building Materials*, 2017. **132**: pp. 170–187.

69. Huseien, G.F., et al., Alkali-activated mortars blended with glass bottle waste nano powder: environmental benefit and sustainability. *Journal of Cleaner Production*, 2020. **243**: pp. 118636.

70. Balapour, M., A. Joshaghani, and F. Althoey, Nano-SiO_2 contribution to mechanical, durability, fresh and microstructural characteristics of concrete: a review. *Construction and Building Materials*, 2018. **181**: pp. 27–41.

71. Rashad, A.M., A comprehensive overview about the effect of nano-SiO_2 on some properties of traditional cementitious materials and alkali-activated fly ash. *Construction and Building Materials*, 2014. **52**: pp. 437–464.

72. Beigi, M.H., et al., An experimental survey on combined effects of fibers and nanosilica on the mechanical, rheological, and durability properties of self-compacting concrete. *Materials & Design*, 2013. **50**: pp. 1019–1029.

73. Nazari, A. and S. Riahi, Improvement compressive strength of concrete in different curing media by Al_2O_3 nanoparticles. *Materials Science and Engineering: A*, 2011. **528**(3): pp. 1183–1191.

74. Deb, P.S., P.K. Sarker, and S. Barbhuiya, Sorptivity and acid resistance of ambient-cured geopolymer mortars containing nano-silica. *Cement and Concrete Composites*, 2016. **72**: pp. 235–245.

75. Nazari, A. and S. Riahi, RETRACTED: splitting tensile strength of concrete using ground granulated blast furnace slag and SiO_2 nanoparticles as binder. *Energy and Buildings*, 2011. **43**(4): pp. 864–872.

76. Kumar, M., N. Malay, and J. Kujur, Study of natural carbonation of concrete incorporating marble dust. *Proceedings of the Institution of Civil Engineers–Construction Materials*, 2018. **171**(2): pp. 85–92.

77. Kumar, S., A. Kumar, and J. Kujur, Influence of nanosilica on mechanical and durability properties of concrete. *Proceedings of the Institution of Civil Engineers – Structures and Buildings*, 2019. **172**(11): pp. 781–788.

78. Duan, P., et al., Effects of adding nano-TiO_2 on compressive strength, drying shrinkage, carbonation and microstructure of fluidized bed fly ash based geopolymer paste. *Construction and Building Materials*, 2016. **106**: pp. 115–125.

79. Li, G., et al., Effects of nano-SiO_2 and secondary water curing on the carbonation and chloride resistance of autoclaved concrete. *Construction and Building Materials*, 2020. **235**: p. 117465.

80. Kalakada, Z., J.-H. Doh, and S. Chowdhury, Glass powder as replacement of cement for concrete–an investigative study. *European Journal of Environmental and Civil Engineering*, 2019: pp. 1–18.

81. Li, L.G., et al., Synergistic cementing efficiencies of nano-silica and micro-silica in carbonation resistance and sorptivity of concrete. *Journal of Building Engineering*, 2021. **33**: p. 101862.

82. Lim, S. and P. Mondal, Effects of incorporating nanosilica on carbonation of cement paste. *Journal of Materials Science*, 2015. **50**(10): pp. 3531–3540.

83. Hanus, M.J. and A.T. Harris, Nanotechnology innovations for the construction industry. *Progress in Materials Science*, 2013. **58**(7): pp. 1056–1102.

84. Zhou, C., et al., Enhanced mechanical properties of cement paste by hybrid graphene oxide/carbon nanotubes. *Construction and Building Materials*, 2017. **134**: pp. 336–345.

85. Zhang, S.-L., et al., Effect of a novel hybrid TiO_2-graphene composite on enhancing mechanical and durability characteristics of alkali-activated slag mortar. *Construction and Building Materials*, 2021. **275**: pp. 122154.

86. Praveenkumar, T.R., M.M. Vijayalakshmi, and M.S. Meddah, Strengths and durability performances of blended cement concrete with TiO_2 nanoparticles and rice husk ash. *Construction and Building Materials*, 2019. **217**: pp. 343–351.

87. Çevik, A., et al., Effect of nano-silica on the chemical durability and mechanical performance of fly ash based geopolymer concrete. *Ceramics International*, 2018. **44**(11): pp. 12253–12264.

88. Zhang, M.-h. and H. Li, Pore structure and chloride permeability of concrete containing nano-particles for pavement. *Construction and Building Materials*, 2011. **25**(2): pp. 608–616.

89. Mohammed, M.K., A.R. Dawson, and N.H. Thom, Macro/micro-pore structure characteristics and the chloride penetration of self-compacting concrete incorporating different types of filler and mineral admixture. *Construction and Building Materials*, 2014. **72**: pp. 83–93.

90. Wu, L., et al., Influences of multiple factors on the chloride diffusivity of the interfacial transition zone in concrete composites. *Composites Part B: Engineering*, 2020. **199**: p. 108236.

91. Samadi, M., et al., Enhanced performance of nano-palm oil ash-based green mortar against sulphate environment. *Journal of Building Engineering*, 2020. **32**: pp. 101640.

4 Sustainability of Nanomaterial-Based Self-Healing Concrete

4.1 INTRODUCTION

Concrete is vastly applied as one of the most important building components due to its compressive strength, affordability, durability, and availability of its raw materials [1–6]. Plenty of concrete structures, however, deteriorate and degrade over time. Water permeation adversely affects concrete efficiency [7–10], as it leads to micro- and macro-level cracks that form passages for water ingress, fluids with dissolved particles, and acidic gases to flow [11,12]. These elements permeate and influence the reinforcement; ultimately affecting durability. Thus, the environment–concrete interaction dictates its long-term performance [13]. Besides affecting durability, water permeation of exposed concrete structure corrodes the reinforcing steel bars [9]. Invisible cracks that may go unnoticed would increase in number and size as a result of permeation, expansion, and contraction of materials. This highlights the importance of reliable infrastructure maintenance and inspection procedures. Continuous maintenance and inspection is impossible due to the massive infrastructure scales that incur high costs. Besides, it is sometimes difficult to repair the damaged areas in the affected structures [7,12].

Environment-friendly materials with low carbon print are widely sought in the construction industry to substitute materials based on ordinary Portland cements (OPCs) that worsen carbon pollution. The deterioration in concrete from the start of its service demands more OPC and shortens its lifespan. Further repair work is costly and requires intensive labour. As cement materials have low resistance towards harsh environment, the structures have low durability and short service life. Critical structural problems arise due to concrete cracking and expansion [1,2]. Hence, the self-healing technology (smart concrete) has been vastly explored to enhance the durability of concrete structures [3].

Self-healing enhances material durability, particularly for construction in harsh chemical and physical settings. Certain characteristics of materials, such as thermodynamic and kinetic, are protected in the self-healing method that incorporates nanostructures. Nanomaterials possess exceptional functional features and are faster than ordinary materials due to the existence of many interfacial atoms in nanomaterials. The combination of these nanostructures leads to the fabrication of numerous nanosystems, in which certain elements can be embedded to promote self-healing. This self-healing method is simpler than devising a nanosystem that is robust [4]. With the advancement of nanotechnology and the advent of self-healing materials

DOI: 10.1201/9781003196143-4

for fabrication, nanomaterials refer to elements with a particle size below 500 nm. Self-healing materials autonomously recover damages or may need temperature deployment (external stimuli) for non-autonomic recovery [5].

The self-healing concrete research segment is continuously seeking nanomaterials suitable for sustainable development. This chapter is composed of three sections: Section 4.1 describes the significance of self-healing concrete to develop sustainable, environment-friendly, and pollution-free construction. Next, Section 4.2 looks at some self-healing processes that enable efficient, sophisticated, and elegant repair mechanisms. Several self-healing mechanisms, along with their advantages and disadvantages, are discussed. Lastly, Section 4.3 presents self-healing mechanisms as a viable self-recovery solution to corroding, deteriorating, degrading, and cracking active concretes.

4.2 SUSTAINABILITY OF SMART CONCRETE

The self-healing technology promotes smart building materials with low carbon emission and energy-saving aspects to uphold building's sustainability and energy efficiency. Sustainable development is aimed at keeping the ecology on Earth balanced and undisturbed, so as to protect the future with utmost care and support [6]. Being composed of social benefits, economic security, and environmental safety, sustainability protects biodiversity with an ecosystem that is balanced. To date, the industrial players (architects, engineers, scientists, and policymakers) seek a sustainable model that effectively decreases the adverse effects on the ecosystem. Hence, sustainability is synonymous with green environment and is environment friendly [7,8]. Self-healing materials have garnered much attention in reducing degradation, extending service life, and slashing material maintenance cost [9,10]. Besides saving energy, the self-healing method reduces environmental pollution and OPC consumption, while concurrently enhancing concrete sustainability and service life.

4.3 LIFECYCLE ANALYSIS OF SELF-HEALING CONCRETE

The self-healing mechanism has been extensively explored since the past decade to self-repair concrete cracks with the proposal of several self-healing strategies for cementing materials. For instance, the method of lifecycle assessment (LCA) is standardized in ISO 14040-14044 to examine the effects of numerous cradle-to-grave services and products on the environment. The LCA is aimed at decreasing the adverse effects on the environmental by deploying the self-healing concrete mechanism. Some advantages of this self-healing concrete mechanism are lower deterioration rate, repair frequency, and cost, as well as extended service life. These ultimately lead to sustainable environmental due to fewer repairs and use of material resources, low-pollutant emission, and energy consumption, as well as less transportation and traffic congestion due to repair work [11]. For instance, cracks were self-healed when polyurethane precursor encapsulation generated partial barrier against immediate chloride ingress through the cracks. Chloride concentration decreased up to more than 75% at the cracked area after deploying self-healing concrete. Besides, the lifespan of self-healing concrete in the marine structures can be extended up to

60–94 years, in comparison to 7 years for normal concrete. The calculations of LCA revealed crucial environmental advantages at 56%–75% via extension of service life.

4.4 PRODUCTION OF NANOMATERIALS

Nanoparticles yield superior effect as fillers compared to materials with micron size. According to Guterrez [12], all materials may be converted into nanoparticles via chemical treatment or crushing them. Nanoparticle accuracy is determined by the chemical components and the purity of parent material(s). The two schemes for large-scale nanomaterial production are top-down [13] and bottom-up [14] approaches, which are determined by the nanoscale property of a material, its suitability, and the cost [15]. In the top-down scheme, which is mostly employed at industrial level, bulky structures are converted into nanostructures while retaining their chemical or physical behaviour via control at atomic level [16]. Meanwhile, the milling technique in top-down scheme is preferred due to its cost-efficiency, accessibility, ease of modification, its ability to yield nanoparticles in massive volume, and dismissal of intricate machine or chemical reagent. The noted unpredictable superiority and consistency in the top-down approach, nonetheless, can be addressed via a milling method by increasing milling speed, nature of jar, as well as ball number and type – for generating better quality nanoparticles [17].

High-energy ball milling is commonly applied for fabricating a range of nanomaterials, including nano-quasicrystals, nanoparticles, nanocomposites, nanograins, and nanoalloys. This milling method, initiated by John Benjamin (1970), generates oxide particles in nickel super-alloy matrix so that the alloy components become effective for improving mechanical strength and high-thermal structure. The conversion of materials into the sought morphology can be affected by some factors during milling, such as cold weld, fractures, and deformed/distorted plastics. In the milling method, the material is crushed and blended with other particles to yield new phases with varying compositions. The final yield is flakes that are further refined using preferred ball and milling standard.

Most of the concrete nanomaterials (e.g., nanoclay, nano-silica, and nanoalumina) are produced using the bottom-up scheme that deploys molecular/atomic level through self-assembling process known as 'molecular-level processing' or 'molecular-nanotechnology'. These nanomaterials can be used indirectly to generate chemicals and nanomaterials [18]. Nanoparticles produced from the bottom-up approach are then customized via chemical synthesis. When compared to the top-down scheme, the bottom-up approach produces nanomorphology with better reproducibility and uniformity, as well as nanocrystals with perfect molecular/atomic order. Nanomaterials generated using the bottom-up scheme have high chemical reactivity, electronic conductivity, and optical absorption [19]. Besides gaining tinier size and more consistent atom surface with modified surface morphologies/energies, the bottom-up approach is effective in yielding self-cleaning and self-healing nanomaterials with better pigment features, catalytic properties, and sensing capability. Although the bottom-up scheme is cost-inefficient, limited to laboratory setting, and demand chemical synthesis expertise [20], its nanoparticles are excellent for biotechnology, electronic components, and other advance features.

4.5 PRODUCTION OF NANOCONCRETE

The exceptional progress in nanotechnology has initiated nanomaterials (e.g. nano-kaolin, nano-silica, polycarboxylates, and nanoalumina) that remarkably enhance properties of concrete [15,21,22]. A small amount of nanomaterials can enhance mechanical properties, including compressive strength, splitting tensile and flexural strengths of cement pastes [23,24], mortars [21,25], as well as concretes [26]. Pastes, mortars [21,25], and concretes [26] containing nanomaterials displayed better strength than those with OPC, ascribed to better interfacial bonding despite hardened cement paste and aggregates, lower pore density, as well as rapid pozzolanic reaction and cement hydration process. Nanomaterials may also be deployed to enhance the durability of concrete properties and to lower porosity [27,28]. Along with the advancement in concrete technology, self-healing approach is crucial to achieve smart and sustainable concrete.

Nanoconcrete refers to the use of materials with a particle size below 500 nm as part of cement substitute or admixture to produce concrete. Addition of nanoparticles enhances concrete strength, its bulk properties, and model structure. In generating high-density concrete, nanoparticles are used as admixture or partially cement substitute due to refinement of intersection zones in the cementing material. New-fangled nanostructures may be rendered upon modifying the nanoparticles [29–31]. Deficiencies in concrete microstructures include corrosion, voids, and microporosity stemming from alkaline silica reaction that can be eliminated. Nanomaterials have advanced thus far because of their function as new binding agent with a particle size smaller than the OPC, thus enhancing the hydration gel element by providing a solid and neat structure. Adding a blend of fillers and extra chemical reaction in the hydration scheme yields high-performing novel nanoconcrete with high durability.

Nanotechnology in concrete is still in its infancy stage despite the escalating environmental pollution due to OPC usage and the rising demand for ultra-high performance concrete (UHPC). Conventional mix of silica fumes with UHPC promotes high strength and durability in concrete. Unfortunately, costly and limited accessibility of nanomaterials impede the progress of UHPC and encourage the continual use of conventional high strength concrete (HSC). In order to address these drawbacks, a nanomaterial called nano-silica that mimics the characteristics of silica fume is developed and incorporated with UHPC to enable nanotechnology [32]. Some nanoparticles were synthesized using nano-silica to effectively generate concrete [33], nanoalumina [34], titanium oxide nanoparticles [35], carbon nanotube (CNT) [36], and nanopolycarboxylates [37], which are novel in the nanoconcrete segment and have garnered much attention from multiple parties.

4.6 SIGNIFICANCE OF NANOMATERIALS AS SELF-HEALER

Nanomaterials improve the durability, workability, and strength of construction materials, thus affecting cement hydration kinetics and enhancing cement performance. High-performance nanomaterials can be used to produce sustainable concrete via self-healing technology, such as hardening of bi-components. Nevertheless, the little amount of healing materials in microcapsules is insufficient to bind the cement network with the microcapsules. In order to resolve this issue, nanomaterials have been incorporated to design sustainable concrete.

Amalgamation of self-healing components in concrete is trending. 'Self-healing concrete' is that cementing materials self-repair damages via diverse deterioration mechanisms. Nanomaterials used for sustainable concrete have garnered much attention worldwide. Both nanomaterial and self-healing technologies have yielded durable and sustainable concrete [38,39]. Nanomaterials are mostly used in self-healing to prevent corrosion of steel bars in reinforcement concrete. According to Koleva [38], the performance of reinforced concrete can be enhanced by adding nanoscale elements with desired properties (e.g. core-shell polymer micelles or vesicles) in systems based on cement. However, studies on nanomaterial in light of self-healing concrete are in scarcity.

Qian et al. [40] assessed the curing settings in the presence of water, carbon dioxide, and air in both dry and wet states, as well as the impact of nanoclay and water (used as inner water furnishing agent for hydration) on micro-cracks. As a result, the healing extent was superior to the inclusion of nanoclay in producing better cement material mix concentration. No cracks were noted in new areas for all air cured mixtures; signifying exceptional healing. Although there was interior water supply from nanoclay, weak strengthening was observed that made it difficult to relocate final crack areas from those that pre-existed. The poor recovery with air curing was also depicted by Hua [63], whereby the self-healing behaviour exerted by Engineered Cementitious Composite (ECC) with Super Absorbent Polymer (SAP) capsules and water (interior pool for extra hydration) had been effective. On top of that, a range of repair products were recognized across the cracked areas. The noticeable healed cracks were completely absent as they experienced effective self-recovery mechanism. Sufficient moisture/water is vital to serve as ion transportation media and reactant for additional hydration.

4.7 NANO-SILICA-BASED SELF-HEALING CONCRETE

In developing sustainable concrete, nano-silica (SiO_2) has been broadly applied in UHPC. Figure 4.1 shows the field-emission scanning electron microscopy (FSSEM) images of nano-silica generated from micron-sized silica. The strong effects exerted by nano-silica in UHPC can be compared with micro-silica or silica fume in terms of durability, strength, and performance [41–43]. According to Qing and Zenan [42], nano-silica embedded in concrete offers early strength better than silica fume, thus enhancing its workability with the lowest amount of super-plasticizer. Since the particle size of nano-silica serves as ultra-filler within the concrete, the micro-voids become refined and dense to form smart microstructures [44]. Another advantage of nano-silica is the better water to cement ratio control as the strength can be easily customized. Quercia et al. [45] found that adding nano-silica had improved concrete strength and functioned as cement replacement element. About 20%–30% of cement can be reduced when nano-silica is deployed as cement substitute. Some studies assessed the capacity of nano-silica in reacting with $Ca(OH)_2$ present in concrete to produce C–S–H gel in light of self-healing approach [46–49]. The use of nano-silica for the nanoencapsulation method or for self-healing concrete to function as mineral

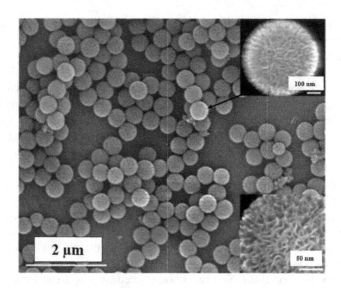

FIGURE 4.1   FSSEM nano-silica image [18].

admixture has been vastly explored. However, nano-silica is costly and is not easily accessible for all areas [18].

4.8   NANOALUMINA-BASED SELF-HEALING CONCRETE

For hydration of cement, alumina and silica are needed to formulate C–(A)–S–H gels with calcium. In cement, alumina controls setting time and silica alters strength. Nanoalumina, yielded from alumina, is infrequently reported for concrete usage despite its capability to accelerate early setting time for high-performance concrete [50,51], thus hindering flocculation and segregation. Disruption in high-performance concrete mixes can lead to cement inhomogeneity which impacts adversely its performance. Apart from its function as a dispersive mediator in cement particles [52,53], nanoalumina refines the porosity in hydration-gel products to turn into nanofiller due to the high fraction of cement in high-performing cement. Grain-size dispersion in concrete is crucial for silica-mediated hydration. Hydration mechanism in the absence of nanoalumina-mediated refinement becomes weak as the silica element fails to penetrate the interior structure of the gel. Embedding nanoalumina forms a path that eases silica or other binding elements to be injected into the hydration gel microstructure interior for refinement [54,55]. The listed advantages indicate nanoalumina as a crucial component in generating sustainable and smart concrete with self-healing method.

4.9   CARBON NANOTUBE-BASED SELF-HEALING CONCRETE

CNTs refer to carbon allotrope with a cylinder-shaped nanostructure. The NTs can be found at a length to diameter ratio up to 132,000,000:1. With typical properties in possession, NTs are applied in multiple nanotechnology and material science domains [12,18]. CNTs may be deployed for numerous structural applications due to their unique electrical, thermal, and conductivity attributes. Belonging to the fullerene family, NTs are incorporated into the hollow and long construction walls in the form of graphene – one-atom-thick carbon sheets that can be rolled at definite orientations with the CNT attributes determined by the revolving radii and angles [36]. The two types of CNTs are single-walled NT (SWNT) and multi-walled NT (MWNT). Each NT is aligned to form 'ropes' that hold the NTs with pi-stacking or van der Waals force [56,57]. The ropes, which reflect chemical bonds in the CNT structure, have the NT bonds resembling graphite and stronger than diamond or alkanes; thus making CNTs incredibly strong.

The flexibility in CNT is crucial to generate rigid or distinctive-type sustainable concrete. The CNT is superior to other nanomaterials for improving strength and flexibility in sustainable concrete [36]. Despite having smaller dimension than other nanomaterials, CNTs improve greatly compressive strength and stress in sustainable concrete. When used in self-healing concrete via ECC, CNT is a promising component for generating smart and sustainable concrete. However, some drawbacks of CNT may compromise its potential in future sustainable construction.

4.10   TITANIUM OXIDE-BASED SELF-HEALING CONCRETE

Titania or titanium oxide (TiO_2) is vastly deployed as pigment in solar cells, food colouring, implants, paints, and photocatalyses. Existing in mineral phase, such as brookite, rutile, and anatase; titania is commonly resourced from anatase, ilmenite, and rutile phases. Titania changes to monoclinic baddeleyite with orthorhombic structures at high pressure, as recently found at the Ries crater in Bavaria [56,58]. The titanium dioxide-containing ore is mostly (98%) found as rutile and ilmenite. When heated to 600°C–800°C, it attains metastable anatase and brookite phases [59].

Including TiO_2 to UHPC or other concretes led to effective self-cleaning capability and served as green element in the construction segment [58]. For instance, TiO_2 was applied by the Jubilee Church in Rome in its product finishing, buildings, and pavements due to its self-cleaning potential [15]. Titanium oxide strengthens concrete by enhancing both its abrasion resistance and performance [60]. Serving as the glassy layer (extreme porosity) or pigment in concrete and UHPC exterior particles, TiO_2 alters their microstructure, and therefore, their performance as well. Protective layers with self-cleaning capability are formed at the concrete surface when the layers mix with hydration gel products. This self-cleaning action in the exterior coats the concrete surface, thus making the extremely durable with permeability. Such feature of TiO_2 can be used in fibre-reinforced system as well to mimic glassy fibre effect. Tailored and refined hydration gel is crucial to provide prolonged durability and strength for concrete sustainability [61]. However, the dusty nature of TiO_2 poses health risks to workers and adversely affects the environment. Besides, TiO_2 generates inflammation and can cause cancer to factory workers [60]. Therefore, any process involving TiO_2 must be carried out cautiously.

4.11 NANOKAOLIN AND NANOCLAY-BASED SELF-HEALING CONCRETE

Kaolinite or nanokaolin refers to the by-product of kaolin – an essential industrial clay mineral; $Al_2Si_2O_5(OH)_4$ [62,63]. This silicate mineral has many layers; a single tetrahedral layer is connected with oxygen atoms to another octahedral alumina layer [64]. China clay or kaolin is rocks with kaolinite [65]. It has white mineral known as dioctahedral phyllosilicate, which is generated by chemical alteration in aluminium silicate (e.g. feldspar) [66]. In ceramic deployment, kaolin undergoes heat treatment and is transformed into $Al_2O_3.2SiO_2.2H_2O$. After treatment or endothermic dehydration crystalline phase, kaolin changes to amorphous structures [67], which indicates a new phase called metakaolin. As it contains amorphous silica and alumina with several hexagonal layers [68], metakaolin is a very reactive pozzolan that can be compared with silica fume. Consistent water penetration is enabled when refinement of metakaolin microstructure enhances durability and strength. Thus, kaolin is cheaper and stronger than silica fume [18,69].

Either bottom-up or top-down approach can be applied to produce nanokaolin, in which the final yield is determined by the processes. Typically, nanokaolin is composed of layers of flakes. Particles of kaolin are similar to nanokaolin, whereby the nanoparticles (after size conversion from microparticles) have wider surface area. Nanokaolin is further treated to generate nanometakaolin when incorporated into concrete to enhance concrete properties [18]. The advantages of using metakaolin in UHPC and other concretes are listed in Ref. [28]. According to Morsy et al. [70], including nanometakaolin into concrete enhanced the compressive strength of mortar by 8%–10%. Nanometakaolin, when compared to OPC, enhanced flexural and tensile strength in mortar by 10%–15% [70,71]. However, several shortcomings were reported regarding the inclusion of nanometakaolin into mortars for UHPC deployment. Besides, poor accessibility of raw kaolin makes it a less popular component than silica fume. Hence, effective guidelines and protocols are sought for large-scale nanometakaolin and nanokaolin production as nanomaterial substitutes to concrete.

Nanoclay, which refers to nanoparticles of layered mineral silicates, can be classified as halloysite, montmorillonite, hectorite, bentonite, and kaolinite based on morphology and chemical composition. Nanoclay, an affordable material with many benefits in polymers, derives from montmorillonite deposits with platelet structures (thickness: 1 nm and width: 70–150 nm). It offers multiple advantages for nanotechnology applications, such as strong chemical reactivity, stability, swelling capacity, interlayer spacing, and elevated hydration. Clays, along with their enhanced organic products, may be assessed to arrive at their chemical compositions using modern and non-intricate equipment, such as powdered X-ray diffraction (PXRD), gravimetric analyser, Fourier transform infrared spectroscopy (FTIR), inductively coupled plasma (ICP), surface area determination, X-ray fluorescence (XRF) spectroscopy, and cation exchange capacity (CEC) using standard ammonium acetate method [58,59. Using cation exchange capacity, the clays can be differentiated in terms of their nature and source. Purity of clay can affect the attributes of nanocomposite. Thus, it is crucial to retrieve montmorillonite with minimum impurity, including kaolin, crystalline silica (quartz), calcite, and amorphous silica [72]. Some methods to purify clays are centrifugation, chemical treatment, hydrocyclone, and sedimentation technique [68].

Despite the affordability and accessibility of clays across the globe and their vast deployment in polymers, their significance as nanomaterials is underreported. Hence, it is vital to examine the material hardness, barrier coating, thermal stability, solvents, as well as its capacity in electronic and other novel segments. Within the context of construction, nanoclay enhances the binding and mechanical attributes of concrete. According to Morsy et al. [70], nanoclay as additive improved tensile and compressive strength in mortar cement. Concrete thermal properties may be improved by incorporating nanoclay as additive in cement paste [73,74]. Qian et al. [40] assessed the addition of nanoclay into water (to promote hydration at micro-cracks) and revealed that the recovery level enhanced substantially after adding nanoclay into the mixtures. The reaction between nanoclay and calcium hydroxide generated C–S–H gel to heal cracks.

4.12 NANOIRON-BASED SELF-HEALING CONCRETE

Copper, cobalt, and nickel refer to ferromagnetic materials with limited applications due to their toxicity and susceptibility to oxidation. Meanwhile, iron oxide nanoparticles have super paramagnetic properties that are beneficial for multiple applications, particularly as colouring and anticorrosion agents in construction materials and coatings. Iron oxide nanoparticles possess exceptional UV blocking capabilities, thus ideal for glass applications; from glass coating to sunglasses. They enable better dispersion in coatings and paints, especially in automotive and high-gloss applications [75,76]. The quick reaction between $Ca(OH)_2$ and Fe_2O_3 nanoparticles generates high amounts of reaction products that can close cracks.

4.13 ECONOMY OF NANOMATERIAL-BASED SELF-HEALING CONCRETES

Concrete is a distinct construction component exploited worldwide with over 4.1 billion tonnes of OPC generated in 2019, for a broad range of applications. The concrete is hoped to attain long service life and tolerance against harsh environment. At the end of their service life, the concrete structures are demolished and are either recycled or thrown away as a waste. It is easier to implement innovation in the building sector than deploying modernization of disruptive goods. Building a complete concrete structure includes skilful trading and products from multiple supplies. A modified structure may be assessed by the construction corporation. Plenty of novel products demand further exploration and approval from construction players prior to deployment at site.

A number of factors should be weighed in prior to nanotechnology incorporation into concretes. First, in handling large-scale concrete materials, safety must be emphasized while constantly being environment-friendly. Next, sufficient field testing is crucial in determining the viability of innovations. Lastly, smart concretes must be better than OPC concretes as it is difficult to demolish the latter structures as they also involve explosives at times to break up the concretes. To address these limitations, nanoparticle deployment with self-healing technology is bound to benefit the functional aspect of the construction domain. The low amount of nanomaterials must not adversely affect material performance and should be in line with the standard construction practices. Smart concrete with self-healing mechanism should be superior to conventional materials, including the control of released admixtures that may penetrate the market.

4.14 SAFETY FEATURES OF NANOMATERIAL-BASED CONCRETES

Carbon dioxide emitted from the process of generating OPC concrete (as a result of de-carbonation of lime-calcination reaction) can be decreased by using nanomaterials embedded with self-healing technology [77]. Apparently, skyscrapers in metropolis and magnificent structures, such as the Taj Mahal, the Sydney Opera House, and the Chrysler Building, all use concretes. In every aspect, durability of concrete is crucial for erection of historical buildings centuries ago in the absence of qualified engineers and cutting-edge technology. Notably, manufacturing smart and new concrete may hinder the wastage of billion tonnes of resources (raw materials) at annual basis in producing concretes, that also in an inefficient manner. Green and sustainable building sought as OPC concretes annually contributes to over 5% of total greenhouse gases emitted into the air across the globe [78]. As such, smart materials that are strong, clean, reliable, safe, and efficient are in need to substitute concretes made of OPC. This coins the following term 'nanomaterial-based smart concrete embedded with self-healing mechanism'.

4.15 SUMMARY

To date, the self-healing mechanism in concrete sustainability has garnered much attention within the context of building domain. Increment in OPC use has severely deteriorated the quality of the environment. The immense benefits of this self-healing technology in concrete have been highlighted from multiple aspects, including environmental safety, building sustenance, and energy-saving features. Many recent studies have looked into the progress, shortcomings, and incoming trends of nanomaterials and self-healing technologies in generating sustainable concrete. The vast literature pertaining to this topic arrived at the gist listed as follows: (i) smart and sustainable concrete with self-healing attribute is characterized by high durability despite harsh condition, lower pollutant release, being eco-friendly, and being cost efficient. (ii) Both the nanotechnology and nanoscience domains are looking at impressive advancement with the advent of nanomaterial-based concrete embedded with self-healing technology, which is viable for future deployments. (iii) Inclusion of nanofibres, nanomaterials, and CNTs not only yields high durability and strength in cementing composites but also ascertains the quality of environment. (iv) The production of materials via nanotechnology route has a crucial role towards future sustainable development, especially within the construction context. (v) Incorporating nanoparticles into concretes is advantageous as it enables the amalgamation of self-healing technology that assures sustainable building. Essentially, this chapter offers a comprehensive taxonomy that outlines the recent research progression within the scope of nanomaterial-embedded concretes with self-healing mechanism.

REFERENCES

1. Jiang, L. and D. Niu, Study of deterioration of concrete exposed to different types of sulfate solutions under drying-wetting cycles. *Construction and Building Materials*, 2016. **117**: pp. 88–98.
2. Chen, Y., et al., Resistance of concrete against combined attack of chloride and sulfate under drying–wetting cycles. *Construction and Building Materials*, 2016. **106**: pp. 650–658.
3. Calvo, J.G., et al., Development of ultra-high performance concretes with self-healing micro/nano-additions. *Construction and Building Materials*, 2017. **138**: pp. 306–315.
4. Gupta, S., S. Dai Pang, and H.W. Kua, Autonomous healing in concrete by bio-based healing agents–a review. *Construction and Building Materials*, 2017. **146**: pp. 419–428.
5. Wang, J., et al., Use of silica gel or polyurethane immobilized bacteria for self-healing concrete. *Construction and Building Materials*, 2012. **26**(1): pp. 532–540.
6. Mastrucci, A., et al., Life cycle assessment of building stocks from urban to transnational scales: a review. *Renewable and Sustainable Energy Reviews*, 2017. **74**: pp. 316–332.
7. Struble, L. and J. Godfrey. How sustainable is concrete. in *International Workshop on Sustainable Development and Concrete Technology*, 2004.
8. Bilodeau, A. and V.M. Malhotra. High-volume fly ash system: the concrete solution for sustainable development. in *CANMET/ACI. Séminaire International*, 2000.
9. Zhu, D.Y., M.Z. Rong, and M.Q. Zhang, Self-healing polymeric materials based on microencapsulated healing agents: from design to preparation. *Progress in Polymer Science*, 2015. **49**: pp. 175–220.
10. He, Z., et al., Facile and cost-effective synthesis of isocyanate microcapsules via polyvinyl alcohol-mediated interfacial polymerization and their application in self-healing materials. *Composites Science and Technology*, 2017. **138**: pp. 15–23.
11. Li, V.C. and E. Herbert, Robust self-healing concrete for sustainable infrastructure. *Journal of Advanced Concrete Technology*, 2012. **10**(6): pp. 207–218.
12. Guterrez, K., How nanotechnology can change the concrete world, 2005.
13. Abdoli, H., et al., Effect of high energy ball milling on compressibility of nanostructured composite powder. *Powder Metallurgy*, 2011. **54**(1): pp. 24–29.
14. Jankowska, E. and W. Zatorski. Emission of nanosize particles in the process of nanoclay blending. in *Quantum, Nano and Micro Technologies, 2009. ICQNM'09. Third International Conference on*, 2009, IEEE.
15. Sanchez, F. and K. Sobolev, Nanotechnology in concrete–a review. *Construction and Building Materials*, 2010. **24**(11): pp. 2060–2071.
16. Shah, S.P., et al., Nanoscale modification of cementitious materials, in Bittnar Z., Bartos P.J.M., Němeček J., Šmilauer V., Zeman J. (eds) *Nanotechnology in Construction 3*. 2009, Springer: Berlin, Heidelberg, pp. 125–130.
17. Saleh, N.J., R.I. Ibrahim, and A.D. Salman, Characterization of nano-silica prepared from local silica sand and its application in cement mortar using optimization technique. *Advanced Powder Technology*, 2015. **26**(4): pp. 1123–1133.
18. Norhasri, M.M., M. Hamidah, and A.M. Fadzil, Applications of using nano material in concrete: a review. *Construction and Building Materials*, 2017. **133**: pp. 91–97.
19. Gesoglu, M., et al., Properties of low binder ultra-high performance cementitious composites: comparison of nanosilica and microsilica. *Construction and Building Materials*, 2016. **102**: pp. 706–713.

20. Paul, K.T., et al., Preparation and characterization of nano structured materials from fly ash: a waste from thermal power stations, by high energy ball milling. *Nanoscale Research Letters*, 2007. **2**(8): p. 397.
21. Li, H., et al., Microstructure of cement mortar with nano-particles. *Composites Part B: Engineering*, 2004. **35**(2): pp. 185–189.
22. Shah, K.W., et al., Aqueous route to facile, efficient and functional silica coating of metal nanoparticles at room temperature. *Nanoscale*, 2014. **6**(19): pp. 11273–11281.
23. Porro, A., et al., Effects of nanosilica additions on cement pastes. in *Applications of Nanotechnology in Concrete Design: Proceedings of the International Conference held at the University of Dundee*, Scotland, UK on 7 July 2005, 2005, Thomas Telford Publishing.
24. Qing, Y., et al., Influence of nano-SiO_2 addition on properties of hardened cement paste as compared with silica fume. *Construction and Building Materials*, 2007. **21**(3): pp. 539–545.
25. Jo, B.-W., et al., Characteristics of cement mortar with nano-SiO_2 particles. *Construction and Building Materials*, 2007. **21**(6): pp. 1351–1355.
26. Schoepfer, J. and A. Maji, An investigation into the effect of silicon dioxide particle size on the strength of concrete. *Special Publication*, 2009. **267**: pp. 45–58.
27. Said, A.M. and M.S. Zeidan, Enhancing the reactivity of normal and fly ash concrete using colloidal nano-silica. *Special Publication*, 2009. **267**: pp. 75–86.
28. Zhang, M.-H. and J. Islam, Use of nano-silica to reduce setting time and increase early strength of concretes with high volumes of fly ash or slag. *Construction and Building Materials*, 2012. **29**: pp. 573–580.
29. Aydın, A.C., V.J. Nasl, and T. Kotan, The synergic influence of nano-silica and carbon nano tube on self-compacting concrete. *Journal of Building Engineering*, 2018. **20**: pp. 467–475.
30. Lim, N.H.A.S., et al., Microstructure and strength properties of mortar containing waste ceramic nanoparticles. *Arabian Journal for Science and Engineering*, 2018: pp. 1–9.
31. Fu, J., et al., Comparison of mechanical properties of CSH and portlandite between nano-indentation experiments and a modelling approach using various simulation techniques. *Composites Part B: Engineering*, 2018. **151**: pp. 127–138.
32. Yu, R., P. Spiesz, and H. Brouwers, Effect of nano-silica on the hydration and microstructure development of Ultra-High Performance Concrete (UHPC) with a low binder amount. *Construction and Building Materials*, 2014. **65**: pp. 140–150.
33. Adak, D., M. Sarkar, and S. Mandal, Effect of nano-silica on strength and durability of fly ash based geopolymer mortar. *Construction and Building Materials*, 2014. **70**: pp. 453–459.
34. Silva, J., et al., Influence of nano-SiO_2 and nano-Al_2O_3 additions on the shear strength and the bending moment capacity of RC beams. *Construction and Building Materials*, 2016. **123**: pp. 35–46.
35. Massa, M.A., et al., Synthesis of new antibacterial composite coating for titanium based on highly ordered nanoporous silica and silver nanoparticles. *Materials Science and Engineering: C*, 2014. **45**: pp. 146–153.
36. Morsy, M., S. Alsayed, and M. Aqel, Hybrid effect of carbon nanotube and nano-clay on physico-mechanical properties of cement mortar. *Construction and Building Materials*, 2011. **25**(1): pp. 145–149.
37. Navarro-Blasco, I., et al., Assessment of the interaction of polycarboxylate superplasticizers in hydrated lime pastes modified with nanosilica or metakaolin as pozzolanic reactives. *Construction and Building Materials*, 2014. **73**: pp. 1–12.
38. Koleva, D., Nano-materials with tailored properties for self healing of corrosion damages in reinforced concrete, IOP self healing materials. The Netherlands, SenterNovem, 2008.

39. Gajanan, K. and S. Tijare, Applications of nanomaterials. *Materials Today: Proceedings*, 2018. **5**(1): pp. 1093–1096.
40. Qian, S., J. Zhou, and E. Schlangen, Influence of curing condition and precracking time on the self-healing behavior of engineered cementitious composites. *Cement and Concrete Composites*, 2010. **32**(9): pp. 686–693.
41. Indumathi, P., S. Shabhudeen, and C. Saraswathy, Synthesis and characterization of nano silica from the Pods of Delonix Regia ash. *International Journal of Advanced Engineering Technology*, 2011. **2**(4): pp. 421–426.
42. Qing, Y., et al., Influence of nano-SiO_2 addition on properties of hardened cement paste as compared with silica fume. *Construction and Building Materials*, 2007. **21**(3): pp. 539–545.
43. Bai, P., et al., A facile route to preparation of high purity nanoporous silica from acid-leached residue of serpentine. *Journal of Nanoscience and Nanotechnology*, 2014. **14**(-9): pp. 6915–6922.
44. Lindgreen, H., et al., Microstructure engineering of Portland cement pastes and mortars through addition of ultrafine layer silicates. *Cement and Concrete Composites*, 2008. **30**(8): pp. 686–699.
45. Quercia, G., G. Hüsken, and H. Brouwers, Water demand of amorphous nano silica and its impact on the workability of cement paste. *Cement and Concrete Research*, 2012. **42**(2): pp. 344–357.
46. Raki, L., et al., Cement and concrete nanoscience and nanotechnology. *Materials*, 2010. **3**(2): pp. 918–942.
47. Lopez-Calvo, H., et al., Compressive strength of HPC containing CNI and fly ash after long-term exposure to a marine environment. *Cement and Concrete Composites*, 2012. **34**(1): pp. 110–118.
48. Grinys, A., V. Bocullo, and A. Gumuliauskas, Research of alkali silica reaction in concrete with active mineral additives. *Journal of Sustainable Architecture and Civil Engineering*. 6(1): pp. 34–41.
49. Wang, L., et al., Effect of nano-SiO_2 on the hydration and microstructure of Portland cement. *Nanomaterials*, 2016. **6**(12): p. 241.
50. Wu, H., et al., Modification of properties of reinforced concrete through nanoalumina electrokinetic treatment. *Construction and Building Materials*, 2016. **126**: pp. 857–867.
51. Sikora, P., M. Abd Elrahman, and D. Stephan, The influence of nanomaterials on the thermal resistance of cement-based composites—a review. *Nanomaterials*, 2018. **8**(7): p. 465.
52. Nazari, A. and S. Riahi, Improvement compressive strength of concrete in different curing media by Al_2O_3 nanoparticles. *Materials Science and Engineering: A*, 2011. **528**(3): pp. 1183–1191.
53. Hosseini, P., et al., Effect of nano-particles and aminosilane interaction on the performances of cement-based composites: an experimental study. *Construction and Building Materials*, 2014. **66**: pp. 113–124.
54. Rosenqvist, J., Surface chemistry of Al and Si (hydr) oxides, with emphasis on nano-sized gibbsite (α-Al $(OH)_3$), 2002.
55. Richardson, I., The nature of CSH in hardened cements. *Cement and Concrete Research*, 1999. **29**(8): pp. 1131–1147.
56. Sobolev, K., et al., Nanomaterials and nanotechnology for high-performance cement composites. *Proceedings of ACI Session on Nanotechnology of Concrete: Recent Developments and Future Perspectives*, 2006: pp. 91–118.
57. Lai, F., M. Zain, and M. Jamil. Nano cement additives (NCA) development for OPC strength enhancer and Carbon Neutral cement production. in *Proceedings the 35th Conference on Our World in Concrete and Structures*, Singapore, 25–27 August 2010, 2010.

58. Maravelaki-Kalaitzaki, P., et al., Physico-chemical and mechanical characterization of hydraulic mortars containing nano-titania for restoration applications. *Cement and Concrete Composites*, 2013. **36**: pp. 33–41.
59. Vallee, F. Cementitious materials for self-cleaning and de-polluting facade surfaces. in *RILEM International Symposium on Environment-Conscious Materials and Systems for Sustainable Development*, 2004, RILEM Publications SARL.
60. Pacheco-Torgal, F. and S. Jalali, Nanotechnology: advantages and drawbacks in the field of construction and building materials. *Construction and Building Materials*, 2011. **25**(2): pp. 582–590.
61. Chen, J. and C.-s. Poon, Photocatalytic construction and building materials: from fundamentals to applications. *Building and Environment*, 2009. **44**(9): pp. 1899–1906.
62. Liu, Q., Y. Zhang, and H. Xu, Properties of vulcanized rubber nanocomposites filled with nanokaolin and precipitated silica. *Applied Clay Science*, 2008. **42**(1–2): pp. 232–237.
63. Bessa, M.J., et al., Moving into advanced nanomaterials. Toxicity of rutile TiO_2 nanoparticles immobilized in nanokaolin nanocomposites on HepG2 cell line. *Toxicology and Applied Pharmacology*, 2017. **316**: pp. 114–122.
64. Adamis, Z. and R.B. Williams, Bentonite, kaolin and selected clay minerals, 2005.
65. Sabir, B., S. Wild, and J. Bai, Metakaolin and calcined clays as pozzolans for concrete: a review. *Cement and Concrete Composites*, 2001. **23**(6): pp. 441–454.
66. Chakchouk, A., et al., Formulation of blended cement: effect of process variables on clay pozzolanic activity. *Construction and Building Materials*, 2009. **23**(3): pp. 1365–1373.
67. Blanchart, P., S. Deniel, and N. Tessier-Doyen, Clay structural transformations during firing. *Advances in Science and Technology*, 2010. **68**: pp. 31–37.
68. Zhang, D., et al., Synthesis of clay minerals. *Applied Clay Science*, 2010. **50**(1): pp. 1–11.
69. Ghafari, E., H. Costa, and E. Júlio, Critical review on eco-efficient ultra high performance concrete enhanced with nano-materials. *Construction and Building Materials*, 2015. **101**: pp. 201–208.
70. Morsy, M., S. Alsayed, and M. Aqel, Effect of nano-clay on mechanical properties and microstructure of ordinary Portland cement mortar. *International Journal of Civil & Environmental Engineering IJCEE-IJENS*, 2010. **10**(01): pp. 23–27.
71. Morsy, M., et al., Behavior of blended cement mortars containing nano-metakaolin at elevated temperatures. *Construction and Building materials*, 2012. **35**: pp. 900–905.
72. Zbik, M. and R.S.C. Smart, Nanomorphology of kaolinites: comparative SEM and AFM studies. *Clays and Clay minerals*, 1998. **46**(2): pp. 153–160.
73. Al-Salami, A., H. Shoukry, and M. Morsy, Thermo-mechanical characteristics of blended white cement pastes containing ultrafine nano clays. *International Journal of Green Nanotechnology*, 2012. **4**(4): pp. 516–527.
74. Heikal, M., et al., Behavior of composite cement pastes containing silica nano-particles at elevated temperature. *Construction and Building Materials*, 2014. **70**: pp. 339–350.
75. Rattan, A., P. Sachdeva, and A. Chaudhary, Use of nanomaterials in concrete, 2016: pp. 81–84.
76. Olar, R., Nanomaterials and nanotechnologies for civil engineering. *Buletinul Institutului Politehnic din Iasi. Sectia Constructii, Arhitectura*, 2011. **57**(4): p. 109.
77. Bondar, D., Alkali activation of Iranian natural pozzolans for producing geopolymer cement and concrete, 2009, The University of Sheffield.
78. Kambic, M. and J. Hammaker, Geopolymer concrete: the future of green building materials, 2012: pp. 1–7.

5 Engineering Properties of High-Performance Alkali-Activated Mortars

5.1 INTRODUCTION

Continued growth in the urban development projects worldwide has made ordinary Portland cement (OPC) increasingly in demand. As of now, OPC remains the principal binding agent in the concrete industry [1,2]. However, OPC is environmentally hostile because its manufacturing process releases high levels of carbon dioxide (CO_2) emission. For every tonne of OPC production, one tonne of CO_2 is produced (a typical ratio of 1:1). Thus, OPC is classified as environmentally harmful binding material [3–5] as CO_2 is considered to be a significant factor in the production of greenhouse gases that cause climate change [6]. Due to industrial and urban development universally, the standards of living have undoubtedly improved. Meanwhile, both industrial and domestic waste management has increasingly become a serious concern worldwide [6,7]. Global collaboration for achieving greater efficiency in waste management, especially to recycle and repurpose the waste resources [8–11], has been noted. Such concerns have enforced the exploration of alternative options in terms of developing new environment-friendly construction materials such as 'green' concrete as well as other products remanufactured from recycled wastes dumped in landfill [12–14].

Lately the alkali-activated mortars (AAMs) and concretes have been introduced as cement-free materials. Generally, such mortars and concretes are prepared involving starting source materials rich in silicon (Si), aluminium (Al), and calcium (Ca) with alkali activation [15–17]. The starting resource materials include meta-kaolin (MK), fly ash (FA), palm oil fuel ash (POFA), ground blast furnace slag (GBFS), and ceramic wastes (WCP) [18–20]. Literature study on AAMs showed excellent properties such as fast setting time, curing at ambient temperatures [21], high early strength [22], high resistance to elevated temperatures, good durability in aggressive environments [12], low CO_2 emission, and energy consumption [23,24]. The alkaline activator solutions normally include sodium hydroxide (NaOH) and sodium silicate (Na_2SiO_3) for the production of AAMs [25,26]. Most of the research studies [22,27,28] revealed that an elevated concentration of NaOH (10–16 M) and high ratio of Na_2SiO_3 to NaOH (2.5) are preferred for the production of high-performance AAMs. Na_2SiO_3 is known to impact negatively on the environment. Besides additional cost, high concentration of NaOH has negative effect on the environment and remains hazardous to the workers [29]. High molarity of NaOH and enriched Na_2SiO_3 in the alkaline solution content are the major problems for the usage of

DOI: 10.1201/9781003196143-5

AAMs as new construction materials. This is a serious concern for the environmental safety because AAM is a mineral-based material that demands great amount of Na_2SiO_3 during synthesis. These deficiencies caused by alkaline solution limit the diversified applications of AAMs in the construction industry. The inclusion Ca from waste materials such as GBFS in alkali-activated mix led to enhance the strength even using low molarity (4 M) of NaOH [30,31]. The compatible nature of C–(A)–S–H and N–A–S–H gels has significant influence on the AAMs and alkaline solution-activated aluminosilicate systems, wherein both products may be obtained [32].

Despite the use of different source materials to prepare AAMs as binders, FA as industrial waste in particular is immensely attractive for the synthesis of AAMs [33–35]. FA is a by-product of coal burning in thermal power plants to produce electricity which contains an ample quantity of amorphous alumina and silica [4]. Therefore, FA's chemical composition renders it a suitable resource material for producing alkali-activated binder. The properties of FA-based AAMs have been examined by several researchers. Due to their excellent durability properties, these AAMs are used as potential cementitious material. Many reports have acknowledged [12,36,37] similar engineering properties of AAMs making them favourable for various construction applications. Yet, several problems exist in using FA-based AAMs for high-curing temperature requirement (40°C–85°C), slow setting time, and low compressive strength. To overcome these drawbacks, one of the waste materials called GBFS has been introduced for enhancing the FA-based AAM properties [28]. The major reaction products of alkali-activated cements for GBFS and FA are calcium silicate hydrate (C–S–H) and amorphous hydrated alkali alumina silicate [22], respectively. Alkali-activated GBFS has high strength, but issues related to rapid setting and insufficient workability along with high values of dry shrinkage limit its applicability [12]. Incorporation of GBFS in FA-based AAMs could enhance the workability, setting time, and strength together with the reduction of solution demand. Otherwise, GBFS has negative effects on mortar's durability exposed to sulphuric acid and sulphate attacks. On top, the high content of CaO in GBFS is the main setback for the reduced resistance of mortar to aggressive environments. However, increasing the GBFS content in alkali-activated matrix can increase the cost, energy consumption, and CO_2 emissions of AAM mixtures.

Glasses are significant waste products that can potentially be used in concrete production, a strategy which could be considered environment-friendly. Several million tonnes of glass bottles are discarded annually worldwide [38,39]. Some of these wastes have already been recycled by glass manufacturers. Nevertheless, it is impossible to recycle all glass wastes due to the variation in colour, imperfections in glass, and processing costs. The potential for using glass in the production of concrete was first investigated some time ago [39,40]. It has been established that pulverized glass wastes derived from bottles could contain significant amount of Al and Si apparently in non-crystalline form. These properties make glass waste a prospective pozzolanic or cement-like substance. Hence, its usage as aggregates in cement production offers a suitable alternative for cement itself. However, it is worth noting that such use could change the characteristics of the finished products [41].

Contemporary concrete production techniques place more emphasis on the inclusion of nanomaterials that offer enhanced features irrespective of whether it is fresh or cured. Most commonly used nanomaterials for performance enhancement of concretes and cement include nanoparticles, such as TiO_2, SiO_2, Al_2O_3, Fe_2O_3, carbon nanotubes/fibres and nano-silica owing to their pore-filling ability and favourable pozzolanic reaction [42–44]. Meanwhile, steady increase in the demand of hybrid cement-like materials (extremely strong, durable, eco-friendly, and sustainable) has been driven by the rapid development of infrastructure worldwide. The performance characteristics of even high-grade concrete can appreciably be enhanced by the addition of nano-silica. In many cases, replacing concrete with as little as 6% nano-silica may achieve significant performance improvements [44]. It is the hydration process that is primarily modified by the introduction of nano-silica, which becomes effectively accelerated. This in turn leads to the generation of a greater quantity of calcium silicate hydrates through the reaction of nano-silica with calcium hydroxide within the concrete and hence the enhanced mechanical characteristics of the end product.

The nano-silica-impregnated concrete contains fewer calcium hydroxide crystals, which cause the formation of more compacted microstructures [45,46]. Approximately 3% increase in the degree of pozzolanic reaction rate has been registered due to the addition of nano-silica in concrete [47]. Besides, a significant increase in the denseness, durability, tensile strength, compressive strength, bending strength, and abrasive resistance has also been recorded in concrete impregnated with nano-silica [47,48]. The permeability of concrete is also affected by the inclusion of nano-silica, with reduced permeability and capillary absorption [44]. Also, concrete containing nano-silica and ground GBFS disclosed enhanced splitting tensile strength (STS) and hydration speed [47].

In this chapter, the influence of nano glass powder content as GBFS replacement on fly ash-based alkali-activated mortars workability, mechanical, and microstructure performance was evaluated. The flow, viscosity, and setting times were considered to assess the performance of proposed mortars. To investigate the mechanical properties, several tests were taken into consideration to evaluate the mortars' performance which included compressive strength, STS, and flexural strength (FS). The X-ray diffraction (XRD), scanning electron microscope (SEM), energy-dispersive X-ray (EDX), Fourier transform infrared spectroscopy (FTIR), thermogravimetric analysis (TGA), and derivative thermogravimetry (DTG) tests were adopted to characterize the microstructure behaviours and the effects of nano glass powder on proposed mortars to improve the morphology structure, to produce the dense gels, and to reduce the total pores.

5.2 WORKABILITY PERFORMANCE

5.2.1 Flowability

The effect of GBFS replacement by BGWNP on the workability of ternary-blended AAMs is presented in Figure 5.1. The workability of AAMs dropped as the BGWNP content increased. The flow diameter decreased from 15.5 to 15.3, 14.7, 14.2, and 13.5 cm with the increase in nano powder content from 0% to 5%, 10%, 15%, and

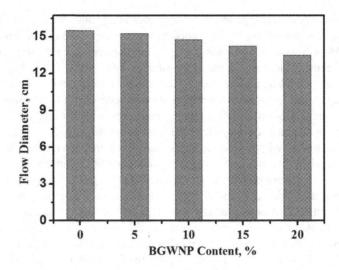

FIGURE 5.1   BGWNP content-dependent flow diameter variations in prepared AAMs.

20%, respectively. The specific surface area of ternary binder was widened with the increase in nano powder content in AAM matrix as GBFS replacement. It was concluded that the high specific area could enhance the water demand, thus leading to reduce workability. This observation is in agreement with the previous findings [48,49]. It was reported [50,51] that the addition of nanomaterials reduces segregation and bleeding and improves cohesiveness of the mixture.

5.2.2 VISCOSITY

The effect of BGWNP replacement on the plastic viscosity of the alkali-activated pastes was obtained according to ASTM D455. The plastic viscosity values were significantly influenced by the content of BGWNP as GBFS replacement. In addition, the plastic viscosity values increased from 88 to 90, 93, 95, and 99 cP with the increase in BGWNP replacement level from 0% to 5%, 10%, 15%, and 20%, respectively. The observed increase in the plastic viscosity of the tested specimens containing high amount of BGWNP could be attributed to their chemical compositions and physical properties' modification. The plastic viscosity was also affected by the speed rate of the setting time of GBFS and high water demand of BGWNP powder. Nath and Sarker [28] reported that the inclusion of GBFS in the geopolymer matrix can accelerate the chemical reaction, increase the viscosity (thus high resistance to the flow), and reduce the flowability of the specimens. Owing to its cementitious properties, the GBFS is classified as the pozzolanic material which could contribute to the hydration process, leading the paste to achieve stronger adhesive properties and lower flowability. According to Diamantonis et al. [52], the appreciable influence of the plastic viscosity by the high content of FA is mainly due to the existence of pores that absorb water without contributing to the higher flowability of the specimen (Figure 5.2).

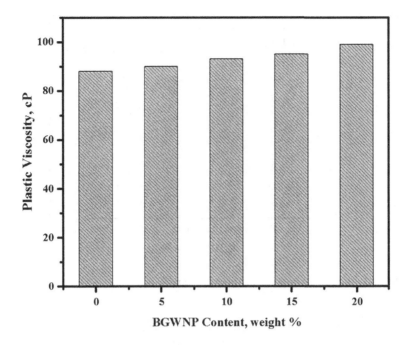

FIGURE 5.2 Influence of BGWNP as GBFS replacement on viscosity of FA-based pastes.

5.2.3 SETTING TIME

Figure 5.3 illustrates the initial and final setting times of prepared AAMs, which were influenced by BGWNP content. Both initial and final setting times were extended with increasing content of BGWNP in AAM matrix. The initial setting time elongated from 34 to 51 minutes with the rise in BGWNP content from 0% to 20% as GBFS replacement, respectively. Similar trend was also observed for final setting time which was extended from 52 to 76 minutes with the increase in BGWNP content from 0% to 20%, respectively. The difference between initial and final setting times was also extended with the increase in BGWNP content in the AAM matrix. This observation suggested that lower the GBFS content in the matrix, lower is the setting rate [28]. Similar trends were reported by others [16,53,54], where a decrease in CaO/SiO$_2$ ratio was found to generate high workability and short setting times. However, the slow rate could enhance the AAM's workability even when the GBFS content reduced from 30% to 10%. This clearly disclosed the significant impact of BGWNP on the AAMs properties. It is reported that nanomaterials such as silica can accelerate the hydration rate and reduce the setting time [49,55].

5.2.4 BULK DENSITY

The bulk density of hardened AAMs was measured using the average weight ratio of three specimens after being surface dried to the volume of specimens. The dry density values of such specimens were obtained at various levels of BGWNP replacing

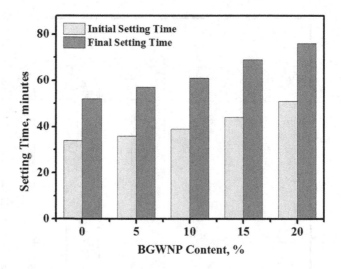

FIGURE 5.3   Initial and final setting times of prepared AAMs.

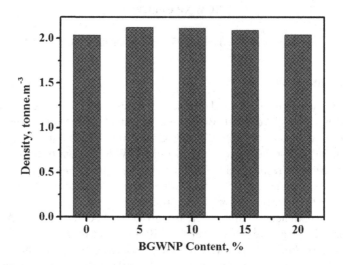

FIGURE 5.4   Effects of BGWNP contents on density of AAMs.

GBFS contents in the ternary blend (Figure 5.4). The density of AAMs increased from 2.04 to 2.12 tonne/m^3 when the content of BGWNP was raised from 0% to 5% to replace GBFS in the matrix. However, the density reduced from 2.12 to 2.04 tonne/m^3 when the BGWNP content increased from 5% to 20%, respectively. The nano powder produced dense gel which led to the increase in density of specimens containing BGWNP compared to control sample which didn't contain BGWNP, that is, 0% BGWNP. Kirgiz [56] reported that the increase of nanomaterials in binder may also decrease the mortars' density. However, as nanomaterial (BGWNP) is increased by 10% of mass of binder, density of mortars decreased over 1.4% higher than that of specimens containing 5% BGWNP content.

5.3 MECHANICAL PROPERTIES

5.3.1 COMPRESSIVE STRENGTH

Figure 5.5 displays the compressive strength of AAMs at ages of 7, 28, 56, and 90 days as a function of BGWNP contents replacing GBFS. At early and late ages, the compressive strength increased monotonically with the increase in BGWNP content from 0% to 5%. At 28 days of curing age, the compressive strength enhanced from 56.2 to 65.5 MPa with the increase in BGWNP content of AAMs from 0% to 5%, respectively. However, the strength reduced to 42.1 MPa when the BGWNP content increased beyond 5% and reached to 20%. Likewise, the specimens prepared with 5% BGWNP replacing GBFS achieved the highest strength compared to others for specimens tested at 56 and 90 days of curing ages. With more benefits reported about the effects of nano-silica in cement- based materials, it is worthwhile to mention that there are some contradicting results related to optimum percentage of nano-silica replacement. It is fair to say that the gain in strength could mainly be attributed to the method of nano-silica production and dispersion of BGWNP in cement-based materials. The main role of pozzolanic reaction was strength development and reduction in pore size distribution. The increase in the compressive strength of mortars

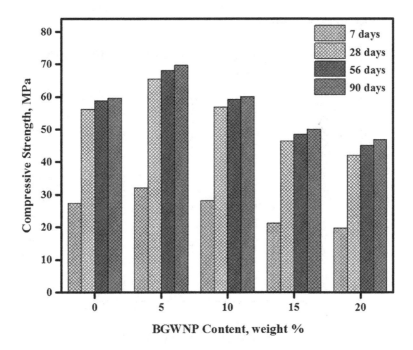

FIGURE 5.5 Effects of varying BGWNP contents on compressive strength development of AAMs.

was noted up to 10% addition of nano-silica and reduced thereafter [46,55]. Some of the studies [42,44] reported that the addition of 4.0%–6.0% of nano-silica by weight of FA could diminish the water absorption and result in denser structure, thereby improving compressive strength of mortars.

5.3.2 Splitting Tensile Strength

Figure 5.6 shows the results on STS of AAMs when GBFS was replaced by BGWNP following 7, 28, 56, and 90-day curing period. For all samples, it was observed that the compressive strength increased with increasing the curing ages from 7 to 90 days. At early age (7 days), the AAM specimens prepared with 5% of BGWNP as GBFS replacement achieved the highest strength (2.21 MPa) compared to control sample and other mortar specimens with BGWNP higher than 5%. For the specimens evaluated at 28 days of curing age, the control samples consisting of GBFS to BGWNP ratio = 30:0 could produce an STS of 3.6 MPa. With greater content (0%–5%, 5%–10%), the STS of AAMs improved (3.6–4.4 MPa, 4.4–4.8 MPa). Moreover, strength loss of 2.9 and 2.8 MPa was observed for AAM samples containing 15% and 20% of BGWP, respectively. Similar trends of results were observed for specimens assessed at 56 and 90 days of curing age. Inclusion of 5% BGWNP led to improve the STS performance. However, rising levels of nanomaterials up to 10% affected negatively on strength performance.

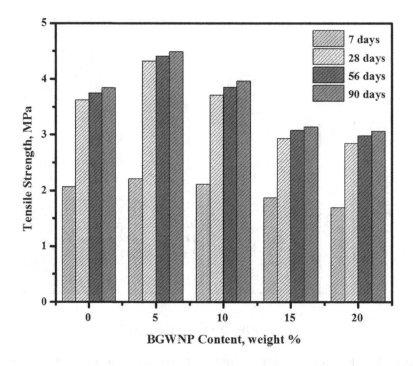

FIGURE 5.6   Influence of AAMs containing BGWNP on tensile strength.

5.3.3 FLEXURAL STRENGTH

Figure 5.7 shows the results obtained from replacing GBFS with BGWNP upon FS of AAM samples. Following the 7, 28, 56, and 90-day curing period, the AAM samples underwent testing for changes in strength characteristics. An inverse relationship in FS was noted in the BGWNP samples when their contents went up to a maximum of 10%. Moreover, there was a noticeable improvement (14%) in FS as BGWNP content increased from 0% to 5%. This increase in FS could be attributed to the microstructure possessing superior hydration characteristics due to the presence of nano powder. Decrease in FS was observed beyond 10% of BGWNP content. First, FS gradually increased (5%–10%, 10%–15%, and 15%–20%) and then decreased (7.2–6.4 MPa, 6.4–5.7 MPa, and 5.7–5.4 MPa). A comparison with the control sample was made at 15% level, where the loss in FS was found to be 9.5%. As mentioned in the literature [57,58], a reduction in strength was found in samples containing BGWNP greater than 10%. That could be attributed to high water demand. Consequently, it produced an adverse effect on the hydration process.

5.3.4 MODULUS OF ELASTICITY

Figure 5.8 shows the effect of BGWNP replacing GBFS on modulus of elasticity (MoE) of AAMs following 28-day curing period. The MoE of AAMs increased with increase in BGWNP levels for GBFS from 0% to 5% and then decreased from 15.4

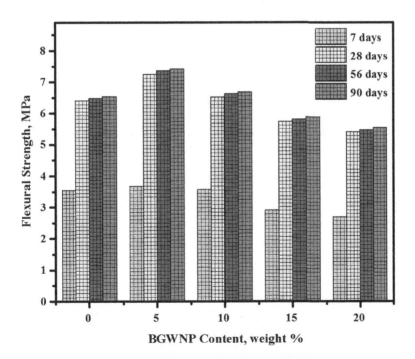

FIGURE 5.7 Influence of AAMs containing BGWNP on flexural strength.

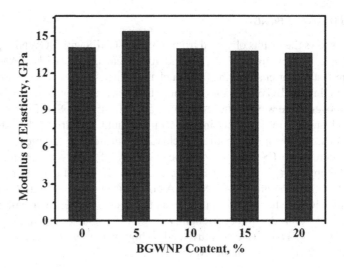

FIGURE 5.8   Influence of AAMs containing BGWNP on MoE.

to 14.4 GPa compared to the control sample (14.1 GPA) containing 0% BGWNP content. Notably, the increased BGWNP contents of AAM samples at 15% and 20% resulted in a respective decrease in the MoE, from 13.8 to 13.6 GPa. The explanation for this was established in the literature [12] where the resulting decrease of FS, STS, and MoE in the AAM samples containing BGWNP levels to a maximum of 10% can be attributed to the mortar's network with reduced level of calcium.

5.3.5   RELATIONSHIP BETWEEN CS, STS, AND FS

Figure 5.9 displays the relationship between splitting tensile, FS, and compressive strength of proposed mortars. Liner regression method was applied to correlate the experimental data as shown in Equations 5.1 and 5.2, with R^2 values between 0.9839 and 0.9873, which signified good confidence for the relationships.

$$STS = 0.0561CS + 0.4995 \qquad (5.1)$$

$$FS = 0.0962CS + 0.9517 \qquad (5.2)$$

where CS is compressive strength (MPa); STS is splitting tensile strength (MPa); and FS is flexural strength (MPa).

5.3.6   STATISTICAL DATA ANALYSIS

For the specimens tested at 28 days of curing age, the statistical results derived from experimental data are presented in Figure 5.10. These results indicated that an increase in BGWNP content increased the interval difference between the minimum and maximum values of compressive, tensile, and FSs as well as the MoE of AAMs.

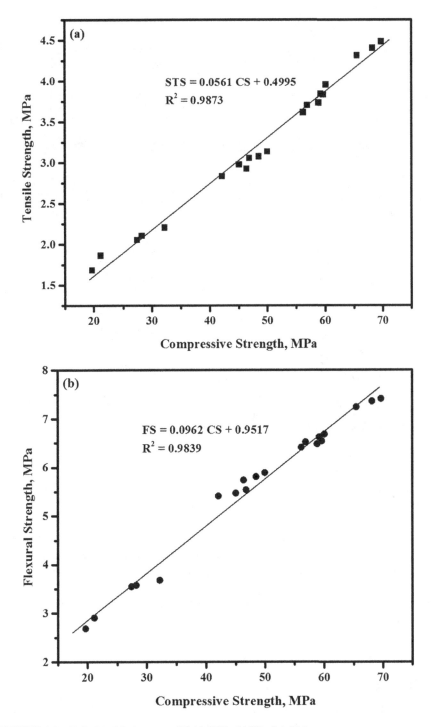

FIGURE 5.9   Relationship between CS (a) STS, (b) FS of AAMs.

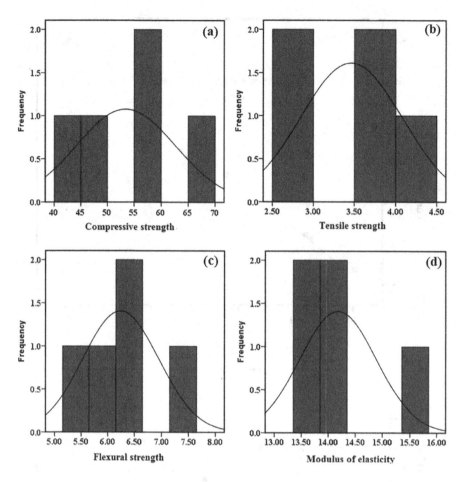

FIGURE 5.10 Histograms of strength properties of AAMs containing various amounts of BGWNP.

Figure 5.10a shows the frequency histogram of compressive strength of AAM specimens. These results depicted that the compressive strength of specimens was normally distributed and fit well with the superimposed normal distribution curve. Similar trends were also observed for the tensile, flexural, and MoE specimens and the frequency histogram displayed a normal distribution as shown in Figure 5.10b–d, respectively.

5.4 MICROSTRUCTURE PROPERTIES

5.4.1 XRD Patterns of AAMs

Figure 5.11 illustrates the XRD patterns and the crystalline structure of the AAMs at curing age of 28 days. An amorphous hallow was identified between 20° and 35°, indicating the presence of AAM gel. The effect of BGWNP inclusion in AAMs

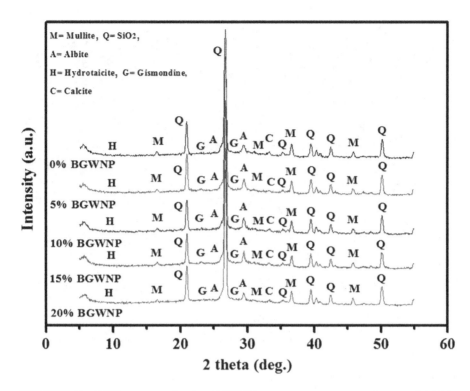

FIGURE 5.11   XRD patterns of various BGWNP contents on structures of prepared AAMs.

matrix appeared in the XRD pattern between 24° and 34° with the increase in intensity of albite ($Na_{0.95}Ca_{0.05}Al_{1.05}Si_{2.95}O_8$) and gismondine ($CaAl_2Si_2O_8 \cdot 4(H_2O)$) peaks. However, the intensity of quartz (SiO_2) peak at 36° decreased with the inclusion of BGWNP in AAMs. The increment in albite and gismondine and reduction in quartz peaks suggested the formation of more C, N–(A)–S–H gels which contributed in the enhancement of hydration and geopolymerization processes. Compared to control sample, the highest intensity of albite and gismondine peaks was observed with 5% and 10% of BGWNP replacing GBFS. On the other hand, the increase in replacement level to 15% and 20% affected the geopolymerization process and restricted the production of C, N–(A)–S–H gels in AAMs, thus leading to lower strength.

5.4.2   FESEM ANALYSES

The impact of BGWNP replacing GBFS in AAMs is examined using microstructural image analyses. Figure 5.12 shows the FESEM images of prepared mortars. Specimens prepared with 5% and 10% of BGWNP exhibited high-performance structure with less pores and non-reacted particles (Figure 5.12a and b). FESEM images manifested the partial reaction of FA particles. The proportion of unreacted FA spheres remaining in the matrix of AAM mixtures cannot be ignored. Apart from the amorphous reaction products, a partial crystalline structure with diameters between 150 and 300 nm was evident. Some needle-shaped crystals on the

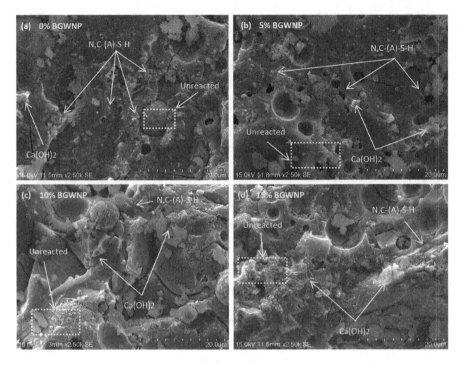

FIGURE 5.12 FESEM images of AAMs containing different proportions of BGWNP (a) 0%, (b) 5%, (c) 10%, and (d) 15%.

surface and surroundings FA particles were seen where some of these crystals are almost covered with a dense layer of amorphous gel. The presence of such crystals was also reported by Jang et al. [59]. In general, the FESEM images depicted that the AAMs were the real mixture of unreacted fly ash particles, some crystals, and a reacted gel phase. With increased level of BGWNP (15% and 20%), more non-reacted fly ash particles appeared with low dense structure as shown in Figure 5.12c and d.

5.4.3 EDX Analyses

The results of EDX spectra of AAMs prepared with 5% and 15% BGWNP are presented in Figures 5.13 and 5.14, respectively. EDX spectra showed that 1.15 ratio of calcium oxide to silica ($CaO:SiO_2$) was high in specimens prepared with 5% of BGWNP, albeit, the addition of BGWNP contributed to a further decrease of CaO in the matrix. However, the AAM specimens prepared with 15% of BGWNP presented a lower ratio of CaO to SiO_2 (0.64). Furthermore, the ratio of SiO_2 to Al_2O_3 increased from 1.74 to 2.55 with rising levels of BGWNP from 5% to 15% as GBFS replacement. The main difference in samples with 5% BGWNP was that the EDX spectra showed relatively low SiO_2/Al_2O_3 which may indicate that higher Al ions were substituted in the C, N–A–S–H chain. In the case where BGWNP content increased from 5% to 15%, the reduction in compressive strength could be attributed to reduce the

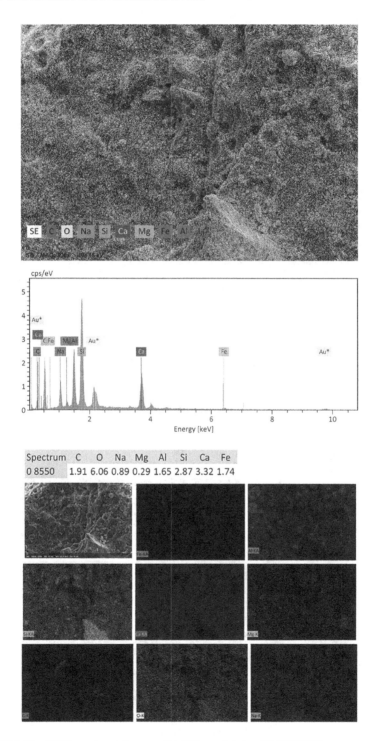

FIGURE 5.13 EDX spectra and maps of AAMs containing 5% BGWNP.

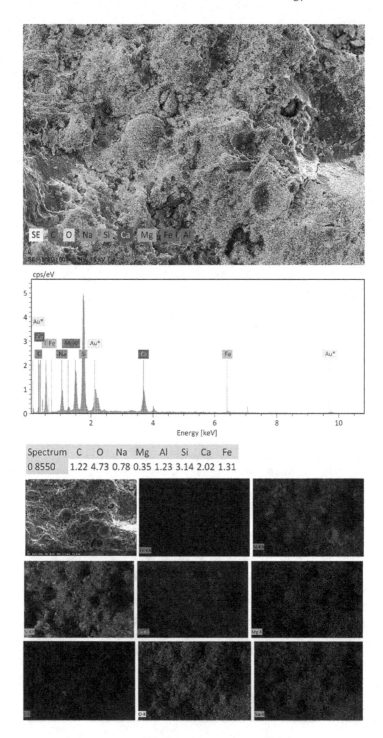

FIGURE 5.14 EDX spectra and maps of AAMs containing 15% BGWNP.

Al_2O_3 and CaO content with increasing amount of SiO_2. This led to the formulation of less gels compared to addition with 5% BGWNP.

5.4.4 FTIR SPECTRA

FTIR measurements were performed to identify the formation of reaction products and degree of geopolymerization in various AAM matrices. Using chemical analysis to find the functional groups based on bonding vibrations, FTIR detected the reaction zones of Si-O and Al-O in AAMs mixtures (Figure 5.15). In such a matrix, development of compressive strength occurred with the dissolution of minerals resulting from the addition of alkaline activators to the base materials. It led to the release of Al via the hydroxylation. This in turn caused the attachment of −OH ions present in the alkali to form Al−O−Al bond by rupturing the weak bonds which liberated a negatively charged Al in IV fold coordination. Finally, a balanced charge was achieved by Ca which reacted preferentially over Na [60]. GBFS containing higher CaO than FA revealed good potential for Ca solubility in the mixture. The quantity of soluble Ca depended on the volume of GBFS present in the mixture which affected directly the compressive strength. It is established that the unit oligomer of (−Si−O−Al) Ca could be built up in chains, sheets, or 3D-framework through the polycondensation process to cause product hardening [61].

The main aim of this study was to replace GBFS by BGWNP to enhance performance strength, durability, and sustainability of AAMs. As the BGWNP content increased

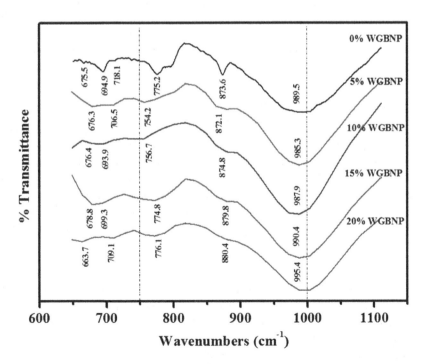

FIGURE 5.15 FTIR spectra of AAMs prepared with different amounts of BGWNP.

from 0% to 5%, the 28 days' strength enhanced from 56.2 to 65.5 MPa. However, the increase in BGWNP contents to 10%, 15%, and 20% allowed the strength to drop to 56.9, 46.4, and 42.1 MPa, respectively. Figure 5.15 depicts such changes. The band of Si–O–Al in 0% BGWNP (989.5 cm^{-1}) then shifted to 985.3 and 987.9 cm^{-1} for AAMs with 5% and 10% BGWNP, respectively. The Si–O–Al band frequency was decreased, indicting an enhancement in C(N)–A–S–H gel product. This led to high homogenous structure of AAMs for 5% and 10% BGWNP content and resulted in higher silicate re-organization compared to 0% BGWNP specimens. With increase in BGWNP levels from 15% to 20%, a lower compressive strength and increase in the band frequency reading to 990.4 and 995.4 cm^{-1}, respectively, were observed. The bending modes of Si-O-Si at 775.2 cm^{-1} were shifted to 754.2 cm^{-1} with increasing BGWNP level from 0% to 5%, respectively. This decrease in the Si-O-Si bond frequency with BGWNP content increase indicated an increment in C–S–H gel product formation. The vibration frequency was decreased with increase in molecular molar mass of the attached atoms. Thus, GBFS released soluble Ca, thereby displacing the Si atoms from Si–O bonds, hence reduction in the vibrational frequency. The addition of BGWNP caused the increment in SiO_2/Al_2O_3 ratio and the vibrational frequency of Si–O–Si (Al) [62].

It was reported that in Ca-based AAMs, the condensation results from hydroxylation of gehlenite and akermanite phases to form cyclo Ca-ortho-sialate-disiloxo (C_3AS_3) and Ca-disiloxonate hydrate (C–S–H), respectively [63]. Another evidence of structural reorganization is the band shifting of AlO_4 vibration from 873.6 to 872.1 and 874.8 cm^{-1} in samples prepared with of 0%, 5%, and 10% FA binder, respectively. Based on these values, increase in BGWNP content was found to cause structural changes in the examined AAMs. Such alterations in AAM network structures could be attributed to the enhancement of C–S–H and C(N)–A–S–H gel formation and increase in nano-silica amount. These changes slowed down the rate of geopolymerization and affected the mechanical strength of AAMs negatively.

5.4.5 TGA and DTG Curves

Figure 5.16 represents the respective TGA and DTG curves of AAMs containing 5% and 15% of BGWNP as GBFS replacement. The TGA and DTG analyses were conducted to determine the weight loss percentage of AAMs. The AAM specimens prepared with 5% of BGWNP as GBFS replacement showed lower weight loss (10.47%) compared to specimens containing 15% of BGWNP which showed more than 12.12% weight loss and revealed stable behaviour. Furthermore, the percentage of C–S–H gel in AAM containing 5% BGWNP was higher than (9.63%) the one calculated with 15% BGWNP (7.82%). The percentage of calcium hydroxide was also calculated. It showed the highest percentage in specimens containing 15% BGWNP (5.73%) compared to 4.53% in AAMs containing 5% BGWNP. The high stability of AAMs containing 5% BGWNP was attributed to the presence of high amount of C–S–H gel and low percentage of Ca (OH)$_2$. This clearly suggests the benefit of BGWNP in enhancing the microstructure properties and increasing the strength performance of proposed AAMs. Singh et al. [64] also acknowledged that the hydration products lead to densification of bulk paste matrix and enhance the mechanical properties as well as the durability of mortars containing nano powder. The existence of calcium

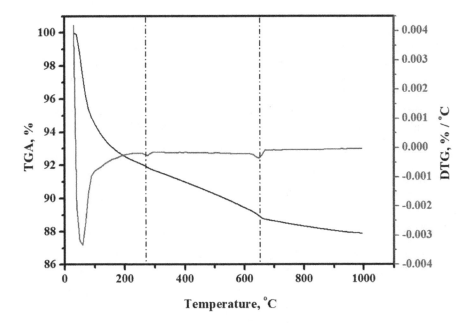

FIGURE 5.16 TGA and DTG curves of prepared AAMs containing BGWNP (a) 5% and (b) 15%.

hydroxide and calcium silicate hydrate in AAMs was calculated using Equations 5.3 and 5.4, respectively:

$$C\text{-}S\text{-}H\,gel(\%) = Total\,LOI - LOI_{CH} - LOI_{CC} \qquad (5.3)$$

where LOI(CH) is the dehydration of calcium hydroxide in the 400°C–550°C range, and LOI(CC) is the carbon dioxide loss in the 600°C–750°C range.

$$CH(\%) = WL_{CH}(\%)\left[MW_{CH}/MW_{H}\right] \qquad (5.4)$$

where WL(CH) is the weight loss ascribed to CH dehydration, MW(CH) is the molecular weights of CH (74 g/mol), and MW(H) is the molecular weight of water (18 g/mol).

5.5 SUMMARY

This chapter disclosed the feasibility of reusing 0%, 5%, 10%, 15%, and 20% of BGWNP as replacement of GBFS to produce AAMs and its effects on workability, mechanical properties, durability, and microstructure behaviour. Based on detailed experimental analyses, the following conclusions emerged:

1. AAM workability increased with increasing BGWNP contents. The initial and final setting times took longer for mixtures containing BGWNP as replacement of GBFS.

2. The achieved compressive strength, FS, STS, and MoE were optimum for AAMs containing 5% BGWNP as replacement of GBFS. By increasing replacement levels up to 10% negatively affected the AAM strength.
3. Inclusion of BGWNP in AAM matrix was highly effective to enhance the durability and to reduce water absorption capacity. AAMs prepared with 5% and 10% of BGWNP showed better performance than the control sample.
4. The XRD, FESEM, EDS, FTIR, and TGA results' analyses showed the enhancement in microstructures and C–S–H gel formation of AAMs containing less than 10% BGWNP.
5. The in-depth XRD, FESEM, and FTIR analyses confirmed the strength improvement of AAMs containing BGWNP activated with low-molarity sodium hydroxide and sodium silicate.

REFERENCES

1. Awal, A.A. and H. Mohammadhosseini, Green concrete production incorporating waste carpet fiber and palm oil fuel ash. *Journal of Cleaner Production*, 2016. **137**: pp. 157–166.
2. Nayaka, R.R., et al., High volume cement replacement by environmental friendly industrial by-product palm oil clinker powder in cement–lime masonry mortar. *Journal of Cleaner Production*, 2018. **190**: pp. 272–284.
3. Deraemaeker, A. and C. Dumoulin, Embedding ultrasonic transducers in concrete: a lifelong monitoring technology. *Construction and Building Materials*, 2019. **194**: pp. 42–50.
4. Huseien, G.F., et al., Geopolymer mortars as sustainable repair material: a comprehensive review. *Renewable and Sustainable Energy Reviews*, 2017. **80**: pp. 54–74.
5. Huseien, G.F., et al., Synthesis and characterization of self-healing mortar with modified strength. *Jurnal Teknologi*, 2015. **76**(1): pp. 195–200.
6. Zhang, P., et al., An integrated gravity-driven ecological bed for wastewater treatment in subtropical regions: process design, performance analysis, and greenhouse gas emissions assessment. *Journal of Cleaner Production*, 2019. **212**: pp. 1143–1153.
7. Sánchez-Escobar, F., D. Coq-Huelva, and J. Sanz-Cañada, Measurement of sustainable intensification by the integrated analysis of energy and economic flows: case study of the olive-oil agricultural system of Estepa, Spain. *Journal of Cleaner Production*, 2018. **201**: pp. 463–470.
8. Esmeray, E. and M. Atış, Utilization of sewage sludge, oven slag and fly ash in clay brick production. *Construction and Building Materials*, 2019. **194**: pp. 110–121.
9. Abdel-Gawwad, H., et al., Recycling of concrete waste to produce ready-mix alkali activated cement. *Ceramics International*, 2018. **44**(6): pp. 7300–7304.
10. Mohammadhosseini, H., et al., Durability performance of green concrete composites containing waste carpet fibers and palm oil fuel ash. *Journal of Cleaner Production*, 2017. **144**: pp. 448–458.
11. Mohammadhosseini, H. and J.M. Yatim, Microstructure and residual properties of green concrete composites incorporating waste carpet fibers and palm oil fuel ash at elevated temperatures. *Journal of Cleaner Production*, 2017. **144**: pp. 8–21.
12. Huseien, G.F., et al., Effects of POFA replaced with FA on durability properties of GBFS included alkali activated mortars. *Construction and Building Materials*, 2018. **175**: pp. 174–186.

13. Huseien, G.F., et al., Waste ceramic powder incorporated alkali activated mortars exposed to elevated temperatures: performance evaluation. *Construction and Building Materials*, 2018. **187**: pp. 307–317.

14. Mohammadhosseini, H., et al., Enhanced performance for aggressive environments of green concrete composites reinforced with waste carpet fibers and palm oil fuel ash. *Journal of Cleaner Production*, 2018. **185**: pp. 252–265.

15. Sun, Z., X. Lin, and A. Vollpracht, Pervious concrete made of alkali activated slag and geopolymers. *Construction and Building Materials*, 2018. **189**: pp. 797–803.

16. Huseien, G.F., et al., Influence of different curing temperatures and alkali activators on properties of GBFS geopolymer mortars containing fly ash and palm-oil fuel ash. *Construction and Building Materials*, 2016. **125**: pp. 1229–1240.

17. Huseiena, G.F., et al., Potential use coconut milk as alternative to alkali solution for geopolymer production. *Jurnal Teknologi*, 2016. **78**(11): pp. 133–139.

18. Huseien, G.F., et al., Compressive strength and microstructure of assorted wastes incorporated geopolymer mortars: effect of solution molarity. *Alexandria Engineering Journal*, 2018. **57**(4): pp. 3375–3386.

19. Kubba, Z., et al., Impact of curing temperatures and alkaline activators on compressive strength and porosity of ternary blended geopolymer mortars. *Case Studies in Construction Materials*, 2018. **9**: p. e00205.

20. Huseiena, G.F., et al., Effect of binder to fine aggregate content on performance of sustainable alkali activated mortars incorporating solid waste materials. *Chemical Engineering*, 2018. **63**: pp. 667–672.

21. Huseiena, G.F., et al., Performance of sustainable alkali activated mortars containing solid waste ceramic powder. *Chemical Engineering*, 2018. **63**: pp. 673–678.

22. Huseien, G.F., et al., Effect of metakaolin replaced granulated blast furnace slag on fresh and early strength properties of geopolymer mortar. *Ain Shams Engineering Journal*, 2016. **9**(4): pp. 1557–1566.

23. Assi, L., et al., Sustainable concrete: building a greener future. *Journal of Cleaner Production*, 2018. **198**: pp. 1641–1651.

24. Turner, L.K. and F.G. Collins, Carbon dioxide equivalent (CO_2-e) emissions: a comparison between geopolymer and OPC cement concrete. *Construction and Building Materials*, 2013. **43**: pp. 125–130.

25. Huseien, G.F., M. Ismail, and J. Mirza, Influence of curing methods and sodium silicate content on compressive strength and microstructure of multi blend geopolymer mortars. *Advanced Science Letters*, 2018. **24**(6): pp. 4218–4222.

26. Huseien, G.F., J. Mirza, and M. Ismail, Effects of high volume ceramic binders on flexural strength of self-compacting geopolymer concrete. *Advanced Science Letters*, 2018. **24**(6): pp. 4097–4101.

27. Salih, M.A., et al., Development of high strength alkali activated binder using palm oil fuel ash and GGBS at ambient temperature. *Construction and Building Materials*, 2015. **93**: pp. 289–300.

28. Nath, P. and P.K. Sarker, Effect of GGBFS on setting, workability and early strength properties of fly ash geopolymer concrete cured in ambient condition. *Construction and Building Materials*, 2014. **66**: pp. 163–171.

29. McLellan, B.C., et al., Costs and carbon emissions for geopolymer pastes in comparison to ordinary Portland cement. *Journal of Cleaner Production*, 2011. **19**(9–10): pp. 1080–1090.

30. Huseien, G.F., et al., Evaluation of alkali-activated mortars containing high volume waste ceramic powder and fly ash replacing GBFS. *Construction and Building Materials*, 2019. **210**: pp. 78–92.

31. Huseien, G.F., et al., Properties of ceramic tile waste based alkali-activated mortars incorporating GBFS and fly ash. *Construction and Building Materials*, 2019. **214**: pp. 355–368.

32. Yip, C.K., G. Lukey, and J. Van Deventer, The coexistence of geopolymeric gel and calcium silicate hydrate at the early stage of alkaline activation. *Cement and Concrete Research*, 2005. **35**(9): pp. 1688–1697.

33. Duan, P., et al., Fresh properties, mechanical strength and microstructure of fly ash geopolymer paste reinforced with sawdust. *Construction and Building Materials*, 2016. **111**: pp. 600–610.

34. Abdollahnejad, Z., et al., Mix design, properties and cost analysis of fly ash-based geopolymer foam. *Construction and Building Materials*, 2015. **80**: pp. 18–30.

35. İlkentapar, S., et al., Influence of duration of heat curing and extra rest period after heat curing on the strength and transport characteristic of alkali activated class F fly ash geopolymer mortar. *Construction and Building Materials*, 2017. **151**: pp. 363–369.

36. Bagheri, A., et al., Alkali activated materials vs geopolymers: role of boron as an eco-friendly replacement. *Construction and Building Materials*, 2017. **146**: pp. 297–302.

37. Al-Majidi, M.H., et al., Development of geopolymer mortar under ambient temperature for in situ applications. *Construction and Building Materials*, 2016. **120**: pp. 198–211.

38. Lin, K.-L., et al., Effect of composition on characteristics of thin film transistor liquid crystal display (TFT-LCD) waste glass-metakaolin-based geopolymers. *Construction and Building Materials*, 2012. **36**: pp. 501–507.

39. Rashad, A.M., Recycled waste glass as fine aggregate replacement in cementitious materials based on Portland cement. *Construction and Building Materials*, 2014. **72**: pp. 340–357.

40. Shi, C. and K. Zheng, A review on the use of waste glasses in the production of cement and concrete. *Resources, Conservation and Recycling*, 2007. **52**(2): pp. 234–247.

41. Zhang, L. and Y. Yue, Influence of waste glass powder usage on the properties of alkali-activated slag mortars based on response surface methodology. *Construction and Building Materials*, 2018. **181**: pp. 527–534.

42. Kaur, M., J. Singh, and M. Kaur, Microstructure and strength development of fly ash-based geopolymer mortar: role of nano-metakaolin. *Construction and Building Materials*, 2018. **190**: pp. 672–679.

43. Sedaghatdoost, A. and K. Behfarnia, Mechanical properties of Portland cement mortar containing multi-walled carbon nanotubes at elevated temperatures. *Construction and Building Materials*, 2018. **176**: pp. 482–489.

44. Adak, D., M. Sarkar, and S. Mandal, Effect of nano-silica on strength and durability of fly ash based geopolymer mortar. *Construction and Building Materials*, 2014. **70**: pp. 453–459.

45. Liu, Y., et al., Effect of a nanoscale viscosity modifier on rheological properties of cement pastes and mechanical properties of mortars. *Construction and Building Materials*, 2018. **190**: pp. 255–264.

46. Björnström, J., et al., Accelerating effects of colloidal nano-silica for beneficial calcium–silicate–hydrate formation in cement. *Chemical Physics Letters*, 2004. **392**(1–3): pp. 242–248.

47. Said, A.M., et al., Properties of concrete incorporating nano-silica. *Construction and Building Materials*, 2012. **36**: pp. 838–844.

48. Nazari, A., et al., Influence of Al_2O_3 nanoparticles on the compressive strength and workability of blended concrete. *Journal of American Science*, 2010. **6**(5): pp. 6–9.

49. Huseien, G.F., K.W. Shah, and A.R.M. Sam, Sustainability of nanomaterials based self-healing concrete: an all-inclusive insight. *Journal of Building Engineering*, 2019. **23**: pp. 155–171.

50. Jalal, M., et al., Mechanical, rheological, durability and microstructural properties of high performance self-compacting concrete containing SiO_2 micro and nanoparticles. *Materials & Design*, 2012. **34**: pp. 389–400.

51. Quercia, G., G. Hüsken, and H. Brouwers, Water demand of amorphous nano silica and its impact on the workability of cement paste. *Cement and Concrete Research*, 2012. **42**(2): pp. 344–357.

52. Diamantonis, N., et al., Investigations about the influence of fine additives on the viscosity of cement paste for self-compacting concrete. *Construction and Building Materials*, 2010. **24**(8): pp. 1518–1522.

53. Phoo-ngernkham, T., et al., High calcium fly ash geopolymer mortar containing Portland cement for use as repair material. *Construction and building materials*, 2015. **98**: pp. 482–488.

54. Huseien, G.F., et al., Synergism between palm oil fuel ash and slag: production of environmental-friendly alkali activated mortars with enhanced properties. *Construction and Building Materials*, 2018. **170**: pp. 235–244.

55. Chithra, S., S.S. Kumar, and K. Chinnaraju, The effect of colloidal nano-silica on workability, mechanical and durability properties of high performance concrete with copper slag as partial fine aggregate. *Construction and Building Materials*, 2016. **113**: pp. 794–804.

56. Kırgız, M.S., Advancements in properties of cements containing pulverised fly ash and nanomaterials by blending and ultrasonication method (review-part II). *Nano Hybrids & Composites*, 2019. **24**(12): pp. 37–44.

57. Yu, R., P. Spiesz, and H. Brouwers, Effect of nano-silica on the hydration and microstructure development of Ultra-High Performance Concrete (UHPC) with a low binder amount. *Construction and Building Materials*, 2014. **65**: pp. 140–150.

58. Thomas, J.J., H.M. Jennings, and J.J. Chen, Influence of nucleation seeding on the hydration mechanisms of tricalcium silicate and cement. *The Journal of Physical Chemistry C*, 2009. **113**(11): pp. 4327–4334.

59. Jang, J.G. and H.-K. Lee, Effect of fly ash characteristics on delayed high-strength development of geopolymers. *Construction and Building Materials*, 2016. **102**: pp. 260–269.

60. García-Lodeiro, I., et al., Effect of calcium additions on N–A–S–H cementitious gels. *Journal of the American Ceramic Society*, 2010. **93**(7): pp. 1934–1940.

61. Yusuf, M.O., et al., Evolution of alkaline activated ground blast furnace slag–ultrafine palm oil fuel ash based concrete. *Materials & Design*, 2014. **55**: pp. 387–393.

62. Lee, W. and J. Van Deventer, The effects of inorganic salt contamination on the strength and durability of geopolymers. *Colloids and Surfaces A: Physicochemical and Engineering Aspects*, 2002. **211**(2): pp. 115–126.

63. Davidovits, J. Application of Ca-based geopolymer with blast furnace slag, a review. in *2nd International Slag Valorisation Symposium*, 2011.

64. Singh, L., et al., Studies on early stage hydration of tricalcium silicate incorporating silica nanoparticles: part I. *Construction and Building Materials*, 2015. **74**: pp. 278–286.

6 Effect of Nanomaterials on Durability Properties of Free Cement Mortars

6.1 INTRODUCTION

The demand of novel construction materials is ever-increasing for environmental sustainable development which can lessen the carbon footprints significantly. The recent increase in the quantities of industrial and agriculture wastes primarily call for effective management of solid waste worldwide and is a major environmental concern. Due to insufficient space for land-filling and increasing cost for land disposal, it has become essential to recycle and utilize industrial waste materials. These include various kinds of industrial by-product waste materials such as fly ash, bottom ash, ceramic, glass, and ground blast furnace slag. Applications of these waste materials in the concrete industry can reduce the disposal concern, making it economical. Natural resources are declining due to its enormous consumption in the construction industry and manufacturing the cement; therefore, it has become crucial to seek for options of replacing this traditional cement with an alternative, which could be used as binder in preparing free cement concrete. Different types of industrial by-products and waste materials that have previously been used in free cement concrete as alternative binder are fly ash (FA), palm oil fuel ash (POFA), waste ceramic (WC), coal bottom ash (CBA), stone dust (SD), ground blast furnace slag (GBFS), and glass wastes (GWs).

In the gentle atmosphere, the ordinary Portland cement (OPC)-based concretes are greatly advantageous in terms of strength and durability performances. In contrast, in the aggressive environments, OPC-based concretes suffer from corrosion because of various acids, sulphate, and chloride attacks [1,2]. These highly reactive chemical compounds are responsible for attack of the concretes and cement pastes, wherein the aggregates/alkali aggregates' reactions are predominant in the aggressive environments. In the OPC and blended cement concretes, the main products from the hydration reaction such as $Ca(OH)_2$ and calcium-silicate-hydrate (C-S-H) gels are responsible for providing the strengths and binding traits of the concretes under the exposure of chemical attacks [3]. This type of chemical attack-based worsening often requires the repair or replacement of the existing concrete structures which are not only very expensive but also labour intensive [4].

It clearly explains the growing need to focus on ways to reduce impact of OPC production on the environment as well as solve durability problems. The problem can be solved by replacing either wholly or partially OPC with other materials that can achieve the object of reducing energy required for production, CO_2 release to the atmosphere, and improve durability. It would further enhance the environmental sustainability when the concrete/binder materials are derived from the industrial and by-product wastes as full or partial replacements to conventional OPC.

Repeated studies verified that the nanoscale materials can act as better super-filler in the porous structures than their micron/bulk counterparts. Guterrez [5] reported that any material may convert into nanoparticles by using chemical treatment or crushing. The chemical constituents and the purity of the source material can decide the accuracy of nanoparticle production. Namely, two routes are followed for the production of large-scale nanomaterials: bottom-up approach [6] and top-down approach [7]. Mainly those approaches have been selected on the basis of cost, expertise, and suitability of the nanomaterials [8]. The top-down milling method is often preferred because of its inexpensiveness, simplicity, availability, and non-requirement of intricate electronics or chemicals. Using this method, it is possible to tailor the physical and chemical traits of the nanomaterials at the atomic levels [9] useful for varied industrial applications. Milling method is advantageous to produce nanoparticles at large scale although the reproducibility, quality, and consistency of the obtained products are not that predictable. However, these limitations of the milling method can be overcome by improving various processing parameters such as ball number and type, speed of milling, and jar type, achieving better nanoparticle morphology [10].

The results of hardened properties of respective alkali-activated mortars (AAMs) are presented in this chapter. The results focused on the durability characteristics of AAM system at different environmental conditions and compared to control sample. The comparisons of durability properties and behaviour include water absorption (WA), resistance to sulphate and acid attacks, carbonation of proposed mortars, resistance to freezing-thawing cycles, abrasion resistance, and the performance of AAMs at elevated temperatures. In addition, microstructural analysis was carried out in terms of scanning electron microscope (SEM) and X-ray diffraction (XRD) to support and better understand the obtained results. With the discussions and results obtained from experimental test data, it clearly shows the effects of replacing GBFS by bottle glass waste nano powder (BGWNP) used in the production of sustainable FA-based AAMs.

6.2   WATER ABSORPTION

Figure 6.1 depicts the WA of AAMs containing BGWNP at ages of 28, 56, and 90 days. The WA decreased with increasing BGWNP to just less than 10%. WA property of AAMs was greatly influenced by the ratio of BGWNP replacing GBFS. The respective values obtained were 8.9, 9.6, 10.4, and 10.6 for 5%, 10%, 15%, and 20% BGWNP content. It can be observed that the specimens containing 5% and 10% of BGWNP showed a packed pore structure which reduced WA compared to control sample (10.2%). However, with the increase in BGWNP level replacing more than 10% GBFS, WA augmented from 10.4% and 10.6% for BGWNP levels of 15% and 20%, respectively. Similar trend of results was also observed for the specimens tested at 56 and 90 days of curing age. The specimens containing 5% BGWNP showed the highest performance compared to other mixtures. The reasons

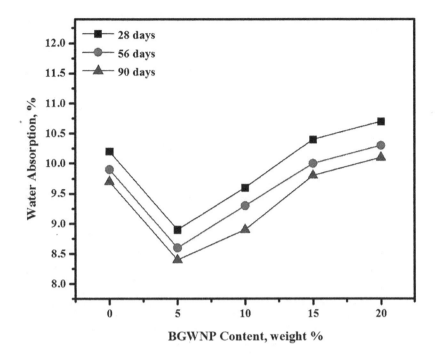

FIGURE 6.1   Water absorption of proposed AAMs.

for such findings can be attributed as follows: as the BGWNP content increased from 0% to 5%, denser C–A–S–H gel was formed which enhanced the homogeneous strength with low WA. Also, the porosity of samples containing 5% BGWNP was reduced to about 12.7% compared to the control sample. This phenomenon could be attributed to the positive effect of BGWNP on the binder's hydration. Generally, the nucleation effect of BGWNP allows the formation of C–S–H phase which is no longer restricted to grain surface alone, thereby enhancing the degree of hydration of binder; also filling more pores by the newly generated C–S–H [11]. For this reason, the porosity of AAMs was found to decrease at first with the addition of BGWNP (5% and 10%). However, since the air content of AAMs increases with the addition of BGWNP, the porosity of BGWNP will increase again when the newly generated C–S–H is insufficient to compensate for the pores that are generated by the entrapped air in fresh mortar. Consequently, considering these two opposite processes, there is an optimal value of BGWNP amount at which the lowest porosity of AAMs can be obtained.

The relationships between the WA and the compressive strength (CS) of AAMs prepared with BGWNP as GBFS replacement are illustrated in Figure 6.2. A reciprocal relation was exhibited between WA and CS. At 28 days of curing age, the WA has dropped from 10.7% to 8.9% as the CS of AAMs increased from 42.1 to 65.5 MPa with the decrease in BGWNP replaced GBFS from 20% to 5%, respectively. Present results are similar to the one obtained by previous researchers [12,13] where the WA decreased with increasing CS of AAMs. The linear regression method was applied to

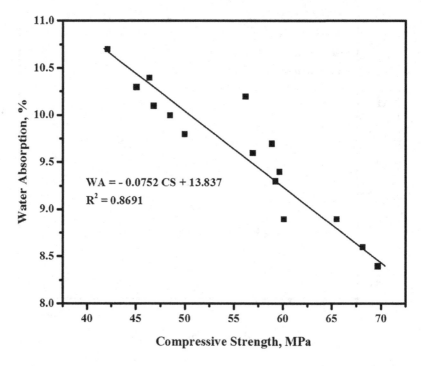

FIGURE 6.2 Correlation between water absorption and compressive strength of AAMs containing BGWNP as replacement of GBFS.

correlate the experimental data as shown in Equation 6.1, with R^2 value of 0.86. This signified good confidence for the relationships.

$$WA = -0.0752CS + 13.837 \qquad (6.1)$$

where WA is the water absorption of AAM specimens and CS is the compressive strength of the tested cube.

6.3   DRYING SHRINKAGE

The mean drying shrinkage values of ternary blended AAMs incorporating FA and GBFS replaced by BGWNP are displayed in Figure 6.3. The rate of shrinkage was high during the early ages up to 28 days which decreased after this age. An inverse relationship was observed between drying shrinkage values and BGWNP content as GBFS replacement. It can clearly be seen from the results that the rate of drying shrinkage was high during the early age's days. The trend in values decreased from 269 to 262, 258, 251, 244 macrostrain with the increasing levels of GBFS replacement by BGWNP from 0% to 5%, 10%, 15%, and 20%, respectively, at 28 days of curing age. Similar trend of results was also recorded after 56 and 90 days of age. At 56 days, the values of drying shrinkage varied between 288 and 254 macrostrain and the specimens containing lower amount of GBFS displayed the lowest value. Finally,

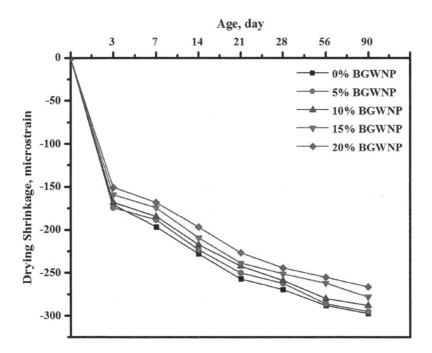

FIGURE 6.3   Effect of BGWNP content on drying shrinkage performance of AAMs.

the 90-day drying shrinkage readings of AAMs with 0%, 5%, 10%, 15%, and 20% BGWNP replacing GBFS discerned to 297, 294, 288, 278, and 266 macrostrain, respectively. The low drying shrinkage values were observed with decreased content of GBFS which could be ascribed to the existence of highly interconnected capillary-like networks inside the alkali-activated concrete matrix. However, GBFS replacement by the BGWNP reduces the CaO content from the mix as the glass has a significantly low calcium content (3.16%). Due to the reduction of CaO content, the hydration rate of alkali-activated binder reduces. As a result, mortars containing high amounts of BGWNP exhibit lower drying shrinkage compared to the control sample concrete (0% BGWNP) [14].

6.4   CARBONATION DEPTH

The carbonation depth readings of AAMs prepared with various levels of GBFS replaced with BGWNP after 90 days of exposure duration are illustrated in Figure 6.4. Three AAM specimens were selected to evaluate the impact of BGWNP content on carbonation depth using phenolphthalein. As compared to AAM specimens prepared with 0% BGWNP, the results showed enhancement in AAM durability with inclusion of the 5% of BGWNP as GBFS replacement and attributed a reduction in carbonation depth to 7.6 mm compared to 8.8 mm recorded with control sample. Several researchers [15–17] reported that the voids ratio was found to reduce with inclusion of the nanomaterials in cement matrix as a result of more dense C–(A)–S–H gels. The reduction in voids content effect positively to decrease

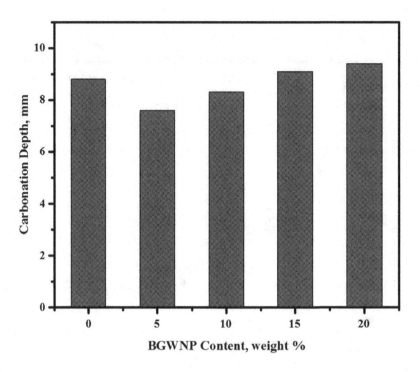

FIGURE 6.4 Carbonation depth of AAMs prepared with various BGWNP as GBFS replacement.

the porosity of proposed mortars led to the reduction in carbonation depth. For all proposed mortar specimens containing BGWNP up to 10%, the increment in carbonation depth was found to be in direct proportionality with nanomaterial's content. The results of measured specimens prepared with BGWNP as GBFS replacement indicated to increase in carbonation depth from 8.3 to 9.1 and 9.4 mm with raising the level of replacement from 10% to 15% and 20%, respectively. In the study by Basheer et al. [18], similar trend of results was observed and it was found that the rate of carbonation within concrete is primarily determined by the water to binder ratio, as well as the porosity and carbon dioxide transport within a mortar matrix. Common knowledge dictates that carbonation propagation is a process of CO_2 diffusion, moving from the environment into the mortar; carbonation depth increases as there is an increase in the diffusion depth of the CO_2. Environmental factors aside, carbonation within mortar/concrete is also an occurring phenomenon due to the pore networks and micro-cracks which exist within concrete specimens and provide routes for CO_2 diffusion.

The relationship between AAMs' WA and carbonation depth is illustrated in Figure 6.5. The measured depth of carbonation was found to be directly proportional to the porosity of proposed mortars [19]. Huseien and Shah [20] reported that the pore structure and high porosity of specimen's impact negatively on the carbonation depth and existence of non-interconnected porous structures that allowed the access of CO_2, leading to diminished emission. The linear regression method (with

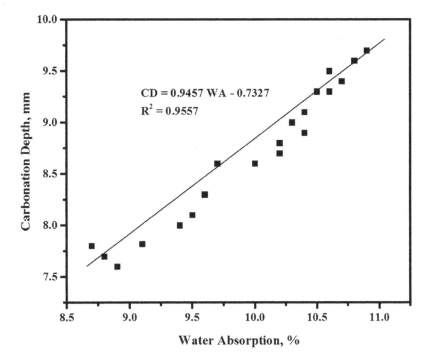

FIGURE 6.5 Relationship between water absorption and carbonation depth of AAMs.

R^2 values 0.95) was applied to correlate the experimental data which signified good confidence following the expression:

$$CD = 0.9457WA - 0.7327 \tag{6.2}$$

where
  CD: carbonation depth of AAM specimens exposed to CO_2 for 90 days, mm
  WA: water absorption of AAMs, %

6.5 ABRASION RESISTANCE

Nowadays, producing sustainable construction materials has been one of the main concerns to researchers in the construction industry. One of the most important durability issues for the hydraulic structures is abrasion erosion (wear), which is considered to determine the service life of hydraulic structures. The abrasion resistance of concrete is the most important property since all the surfaces will interface to the water pressure directly. Abrasion erosion damage is caused by friction and the impact of water-borne silt, sand, gravel, rocks, ice, and other debris on the concrete surface of a hydraulic structure. Figure 6.6 shows the effect of BGWNP as GBFS replacement on abrasion resistance of AAM specimens. For all AAM mixtures, the abrasion resistance improved with the increase in curing age. At different curing ages, the highest abrasion resistance was achieved in specimens containing 5%

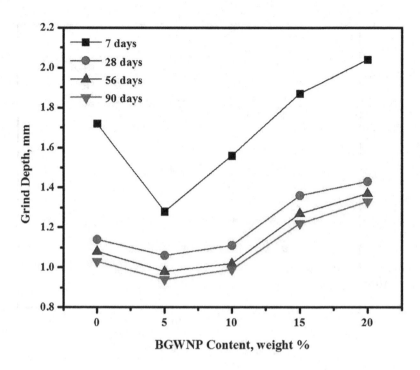

FIGURE 6.6   Abrasion resistance of proposed AAMs containing various levels of BGWNP.

BGWNP as GBFS replacement. The abrasion resistance for the specimens evaluated after 7 days of curing age enhanced by 25.5% and 9.3% with the inclusion of 5% and 10% BGWNP, respectively. However, specimens containing 15% and 20% showed loss in abrasion resistance by 8.7% and 18.6%, respectively. At 28 days of curing age, it was observed that the abrasion resistance improved by 7% by replacing 5% GBFS with BGWNP and the enhancement percentage trend decreased to 2.6% after increasing the replacement level to 10%. For the specimens prepared with up to 10% BGWNP, the results indicated an inverse relationship between abrasion resistance and BGWNP content. It was found that the abrasion resistance decreased by increasing the grind depth with increasing BGWNP-replaced GBFS mortars. The grind depth values increased from 1.11 to 1.36 and 1.43 mm with increasing BGWNP from 10% to 15% and 20%, respectively. Similar trend of results also was observed for the specimens assessed after 56 and 90 days of curing age. The highest resistance was achieved with specimens containing 5% BGWNP as GBFS replacement in AAM matrix.

In a previous study by Liu et al. [21], it was reported that the mortar/concrete with a low porosity, high strength, and strong interfacial bond in the hardened paste could enhance the overall mortar/concrete abrasion erosion resistance performance. As the porosity of the mortar decreases, the mortar becomes more impermeable which increases the abrasion resistance of AAM specimens. The results of CS and porosity were discussed in Sections 5.3.1 and 6.2 that supported the findings of abrasion resistance results in this section. Wang et al. [22] demonstrated the reduction in

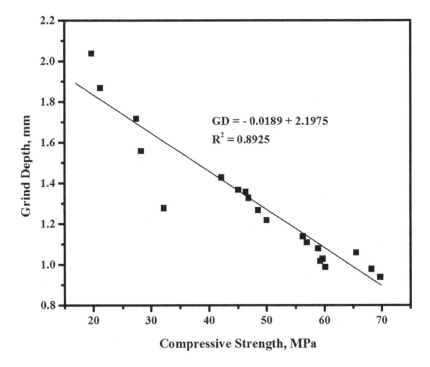

FIGURE 6.7   Relationship between abrasion resistance and compressive strength of AAMs.

CS where the increase in the pore volume in hardened AAM specimens has affected negatively the abrasion resistance.

The grind depth (loss in abrasion resistance) was found to be inversely proportional to the CS of AAM specimens (Figure 6.7). The linear regression method was applied to correlate the experimental data as shown in Equation 6.3, with R^2 value of 0.89. This indicated good confidence for the relationships.

$$GD = -0.0189CS + 2.1975 \qquad (6.3)$$

where
    GD: grind depth for tests AAM specimens, mm
    CS: compressive strength of AAM specimens, MPa

6.6   RESISTANCE TO FREEZE-THAWING CYCLES

Resistance of mortar/concrete to freezing-thawing cycles is very important to measure the durability performance in aggressive environments. In this study, the freezing-thawing resistance of AAMs exposed to 300 freeze/thaw cycles was evaluated at 28 days of curing age. The CS loss, weight loss, change in ultrasonic pulse velocity, and surface deteriorations were evaluated after every 50 freezing-thawing cycles. The influence of BGWNP as GBFS replacement on the freezing-thawing resistance of AAMs was widely assessed to determine mortars' durability. BGWNP content-dependent CS loss, internal frost damage, and surface scaling of AAMs were examined. All prepared mortar specimens were exposed to 300 freezing-thawing cycles at age of 28 days. The influence of BGWNP replacing 0%, 5%, 10%, 15%, and 20% GBFS on the durability of AAM specimens is presented in Figures 6.8–6.11. Figure 6.8 shows the loss in CS. It was observed that the loss in strength increased with increasing number of freezing-thawing cycles for all mortar mixtures. After 300 cycles, it was found that replacing GBFS by 5% BGWNP led to improve the durability performance and reduced the loss in strength from 39.2% to 26.7%. However, by increasing the BGWNP content from 10% to 15% and 20% led to increase the loss in strength from 33.6% to 42.3% and 43.6%, respectively. It can clearly be seen that 5% of BGWNP is the optimum amount to enhance the mortars' resistance to freezing-thawing cycles.

Figure 6.9 shows the weight loss of AAM specimens prepared with various levels of BGWNP as GBFS replacement which were exposed to 300 freezing-thawing cycles at 28 days of curing age. For all mortar specimens, it was observed that the weight loss increased with increasing number of freezing-thawing cycles. The weight loss dropped from 7.2% to 4.3% and 5.9% with increasing levels of GBFS replacement by BGWNP from 0% to 5% and 10%, respectively. On the other hand, the increasing BGWNP contents of 15% and 20% led to reduce the durability performance of AAMs and recorded weight loss to 8.1% and 8.5%, respectively. Compared to control sample, the mortar prepared with 5% of BGWNP displayed the lowest weight loss and showed

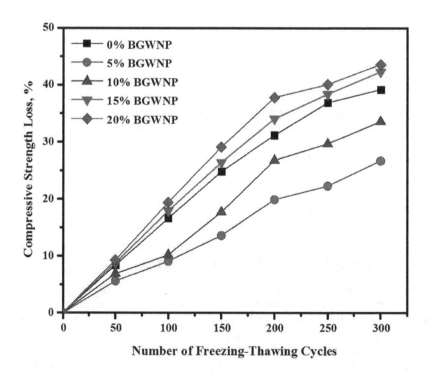

FIGURE 6.8 Strength loss of AAM specimens exposed to 300 freezing-thawing cycles.

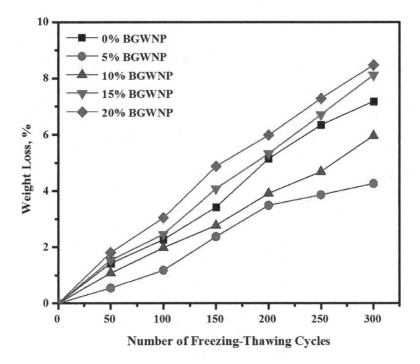

FIGURE 6.9 Weight loss for AAM specimens after 300 freezing-thawing cycles.

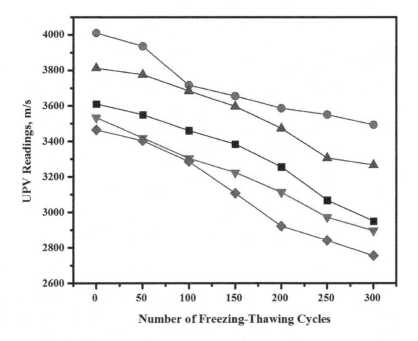

FIGURE 6.10 Impact of BGWNP replacing GBFS on ultrasonic pulse velocity of AAMs.

FIGURE 6.11 Effect of BGWNP content on the surface texture of AAMs.

high durability which can be adopted to design high performance for aggressive environments. The enhancement on mortars' resistance to freezing-thawing cycles with the inclusion of 5% BGWNP as GBFS replacement attributed to improve the hydration process as optimum content of nanomaterial which could have formulated dense surface with lower porous content. As mentioned in Section 6.2, the porosity of AAMs is reduced and the pores are refined with increasing BGWNP content from 0% to 5% as GBFS replacement, both of which contribute to reduced ice formation [23]. Meanwhile, pores become more disconnected leading to the reduction in the capillary transport of external liquid into concrete pores during freezing-thawing cycles. This, in turn, creates less ice growth which could be a major scaling mechanism governed by cryogenic suction of surface liquid under freezing [24].

The ultrasonic pulse velocity (UPV) test was conducted to evaluate the internal cracks and deterioration of AAMs exposed to 50, 100, 150, 200, 250, and 300 of freezing-thawing cycles and the results are depicted in Figure 6.10. For all the AAMs, deterioration increased with the increase in number of freezing-thawing cycles. The UPV readings tend to decrease with increasing exposure time which means that more internal cracks are created due to the formation of ice expansion.

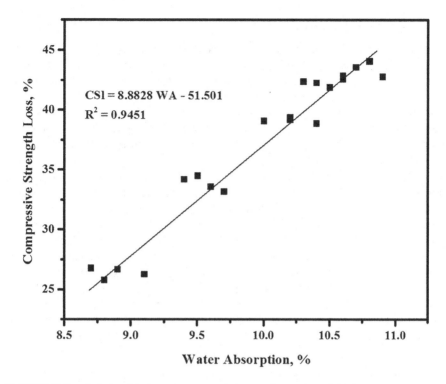

FIGURE 6.12   Relationship between strength loss and porosity of AAMs exposed to 300 freezing-thawing cycles.

The specimens prepared with 5% and 10% BGWNP showed better performance than the control sample. However, the increasing content of nanomaterial up to 10% led to more deterioration and lower resistance to freezing-thawing cycles. With the increase in number of cycles, apparent damages appeared mainly on the surfaces and edges of the specimens as shown in Figure 6.11. Higher amount of deterioration was observed with specimens containing 20% BGWNP as GBFS replacement. It is well known that the increase in silica content with reduced calcium content increased the porosity which could lead to more deterioration. Furthermore, increasing voids in the AAM structure facilitated due to freezing-thawing cycles and destroyed the interlock between particles of mortars and subsequent loss in bond strength [25,26].

Figure 6.12 presented the relationship between resistance to freezing-thawing cycles in terms of CS loss and porosity of AAMs. An increase in the porosity has led to reduce the specimen's resistance to freezing-thawing cycles. A linear relation was established between resistance to freezing-thawing cycles and porosity of mortars with a good R^2 values around 0.96 as follows:

$$CSl = 8.8828 WA - 51.501 \qquad (6.4)$$

where

   CSl: the loss on compressive strength after 300 of freezing-thawing cycles, %
   WA: water absorption of AAMs at 28 days of curing age, %

6.7 ACID ATTACK RESISTANCE

It is well known that the chemical composition of alkali-activated binder affects highly on performance of AAMs in sulphuric acid (H_2SO_4) environment. In this section, the AAMs' resistance against acid attack was determined in terms of CS loss, weight loss, UPV loss, visual appearance, and change on microstructure properties. For this purpose, the AAM specimens were exposed to 10% H_2SO_4 solution for 180 and 365 days to evaluate the effect of BGWNP content on acid resistance of prepared mortars. For all AAMs, it was found that the deterioration tends to decrease with increasing BGWNP contents as GBFS replacement. At 28 days of curing, the AAM specimens were immersed in acid solution with concentration of 10% and were evaluated after 180 and 365 days. Figure 6.13 shows the impact of BGWNP contents on strength loss of AAMs. After 180 days of immersion period, the loss in strength decreased from 5.18% to 3.83%, 3.31%, 2.86%, and 2.41% with increasing replacement levels of GBFS by BGWNP from 0% to 5%, 10%, 15%, and 20%, respectively. Likewise, the durability performance of AAM specimens tested after 365 days exhibited enhancement with increasing content of BGWNP as GBFS replacement. The percentage of strength loss was dropped from 15.1% to 12.4%, 10.1%, 7.8%, and 7.3% with increasing contents of BGWNP from 0% to 5%, 10%, 15%, and 20%, respectively. The inclusion of BGWNP in alkali-activated matrix contributed to improve the specimens' resistance to acid solution and controlled the specimens'

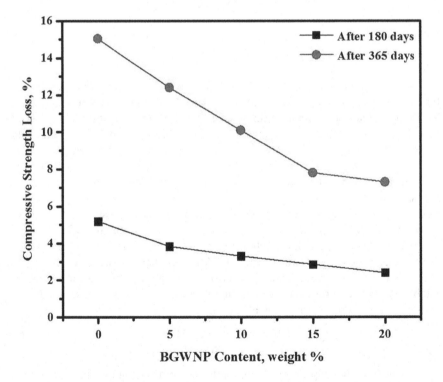

FIGURE 6.13   Strength loss percentage of AAMs exposed to 10% H_2SO_4.

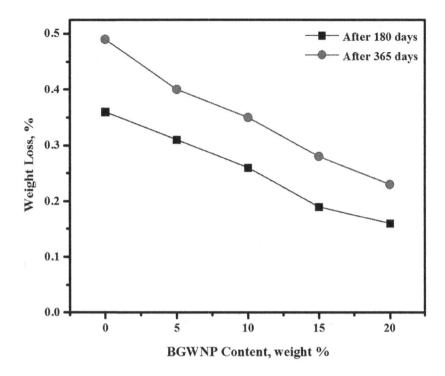

FIGURE 6.14   Effect of BGWNP content on weight loss of AAMs exposed to 10% H_2SO_4.

deterioration. As shown in Figure 6.14, the weight loss percentage decreased slightly from 0.36% to 0.31%, 0.26%, 0.19%, and 0.16% with rising level of replacement GBFS by BGWNP from 0% to 5%, 10%, 15%, and 20%, respectively. After 365 days, the inclusion of 5%, 10%, 15%, and 20% nanomaterials in alkali-activated matrix led to enhance the specimens' resistance and reduced the weight loss to 0.4%, 0.35%, 0.28%, and 0.23% compared to 0.49% for control sample. It could be concluded that the BGWNP reduces the internal cracks (as shown from UPV reading results in Figure 6.15) and improves the durability properties of mortars.

In general, the increased BGWNP content in alkali-activated matrix as GBFS replacement has an effect in reducing the calcium content. Several researchers [27–29] reported that the deterioration can cause the strength loss, expansion, spalling of surface layers, and, finally, disintegration. Most experts attribute sulphate attack to the formation of expansive ettringite ($3CaO \cdot Al_2O_3 \cdot 3CaSO_4 \cdot 32H_2O$) and gypsum [calcium sulphate dehydrate ($CaSO_4 \cdot 2H_2O$)], which may be accompanied by the expansion or softening of prepared mortars. The effect of calcium level on durability performance of AAS in acidic environments was evaluated by several researchers [30,31]. Huseien et al. [30] investigated the effect of slag (GBFS) content (as calcium resource) on acid resistance of ternary blended AAS incorporating waste ceramic powder (WCP) and FA. They reported that the reduction in GBFS content led to enhance the durability performance and reduction in strength loss of tested specimens.

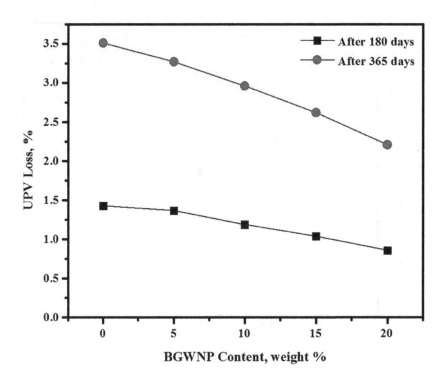

FIGURE 6.15   Effect of acid solution on AAM UPV reading loss.

Figure 6.16 shows the deterioration in AAM specimens exposed to 10% sulph-uric acid solution after 365 days. It was observed that the surface deterioration and the number of cracks and sizes decreased with the increase in BGWNP content as GBFS replacement. It is well known that the nano-silica reduces the amount of cal-cium oxide which could contribute to enhance the mortars' resistance against acid attack by reducing the porosity due to the restriction in calcium hydroxide formula-tion. Upon exposing the AAMs to sulphuric acid, the $Ca(OH)_2$ compound in mortar reacted with SO_4^{-2} ions and formed gypsum ($CaSO_4.2H_2O$). This caused the expan-sion in alkali-activated matrix and additional cracking in the interior of specimens as reflected in the visual appearance results (Figure 6.16). Several researchers [32–35] reported that the reduction in CaO contents reduced the Portlandite and gypsum for-mation, thus increasing the durability of AAMs to sulphuric acid attack.

The effect of 5% BGWNP replaced GBFS (supposed to an optimum amount) on XRD patterns of AAMs exposed to sulphuric acid solution was examined. The results of AAMs exposed to 10% H_2SO_4 for 365 days are presented in Figure 6.17. The samples which carried 0% BGWNP replacement still had clear main phases as well as gypsum and Portlandite. It is quite evident that the prominent XRD peaks matched to quartz (SiO_2), particularly at 2θ values of 26.8°, 40°, and 50° than those obtained before acid exposure. Gypsum ($CaSO_4.2H_2O$) was also found as a new peak at $2\theta = 12.8°$ and 31.2° in the AAMs containing 0% BGWNP. The occurrence of a new peak at 21° was assigned to the gmelinite phase in the mixes. There were less

FIGURE 6.16   Effect of sulphuric acid solution on surface texture of AAMs.

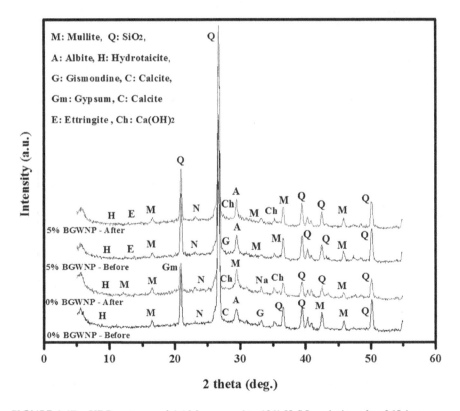

FIGURE 6.17   XRD patterns of AAMs exposed to 10% H_2SO_4 solution after 365 days.

noticeable changes in the peak concentration of samples incorporating 5% BGWNP, before and after they were exposed to 10% H_2SO_4 solution. The XRD findings imply that the BGWNP replacing GBFS increases the mortar's resistance to sulphuric acid attack. There could be a relationship between that and the rate at which AAMs deteriorate in sulphuric acid.

6.8   RESISTANCE TO ELEVATED TEMPERATURES

In this section, the effects of BGWNP as GBFS replacement on resistance of AAMs to elevated temperatures were evaluated. After 28 days of curing age, AAM specimens were subjected to elevated temperatures of 400°C, 700°C, and 900°C and durability performance was measured in terms of CS loss, weight loss, UPV loss, cracks and surface deterioration as well as microstructure properties such as XRD and SEM. For all AAMs, it was found that the degree of deterioration increased with the increase in exposure temperature. Figure 6.18 illustrates the CS loss percentages. It can be clearly seen that the loss in strength decreases with the increase in BGWNP content as GBFS replacement. On the other hand, the strength loss percentages for all AAMs specimens were highly influenced by exposure temperatures and increased with increasing temperatures. The specimens which were exposed to 400°C showed improvement in resistance to elevated temperatures with increasing BGWNP content from 0% to 5%, 10%, 15%, and 20% as the strength loss dropped from 36.1% to

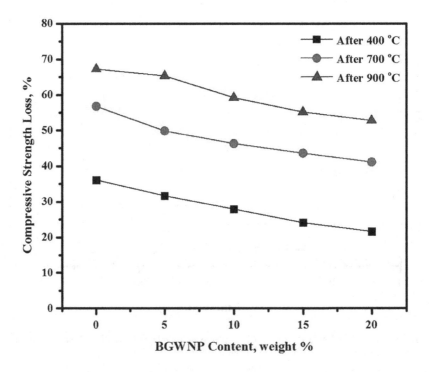

FIGURE 6.18   Strength loss percentage of AAMs exposed to elevated temperatures.

31.7%, 27.9%, 24.1%, and 21.6%, respectively. Similar trend in results was observed for specimens exposed to 700°C temperature and the strength loss decreased from 56.8% to 41.1% with increasing BGWNP levels from 0% to 20%, respectively. The highest strength loss was found with AAM specimens exposed to 900°C which recorded the strength loss of 67.2%, 65.3%, 59.2%, 55.2%, and 52.8% for specimens containing 0%, 5%, 10%, 15%, and 20% BGWNP, respectively.

For the weight loss of AAM specimens exposed to elevated temperatures, it was found to be inversely proportional to BGWNP contents as shown in Figure 6.19. From the obtained results, it can be clearly seen that weight loss increased slightly with the increase in temperatures from 400°C to 900°C. For control sample (without nanomaterials), the increasing temperatures from 400°C to 700°C to 900°C affected the weight loss and increased from 3.4% to 12.8% to 14.1%, respectively. However, the specimens exposed to 900°C showed drop in strength loss from 14.1% to 12.3% to 11.7% to 11.2% to 10.9% with increasing BGWNP levels as GBFS replacement from 0% to 5%, 10%, 15%, and 20%, respectively. Inclusion of BGWNP led to improve the mortars' durability and the reduction in external and internal deterioration as depicted in UPV results (Figure 6.20). The results for UPV exhibited deterioration and the internal cracks increased with the increase of temperatures to 400°C, 700°C, and 900°C. Meanwhile, increasing BGWNP with replaced GBFS in various specimens containing high volumes of FA reduced the internal deterioration. Increased contents of BGWNP in AAMs matrix could have affected in reducing C, (N)–A–S–H and C–S–H gel losses and displayed high performance at

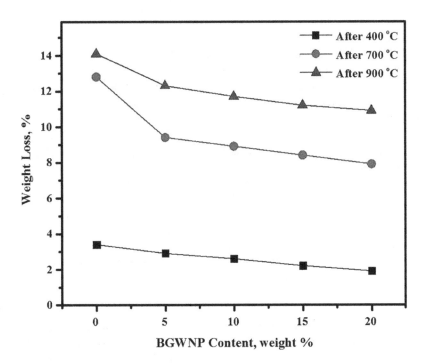

FIGURE 6.19 Effect of elevated temperatures on AAM weight loss percentage.

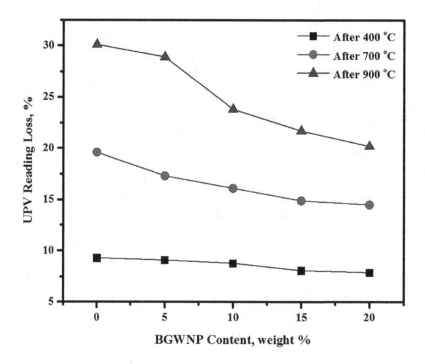

FIGURE 6.20   UPV reading loss percentage of AAMs exposed to elevated temperatures.

elevated temperatures. Conversely, specimens containing 5%, 10%, 15%, and 20% of BGWNP exhibited more stable surfaces at elevated temperatures compared to specimens without BGWNP. This finding was similar to the report by Abdulkareem et al. [36]. Furthermore, the weight loss rate stabilized from 150°C up to 780°C, which could be attributed to the evaporation of both chemically bonded water and hydroxyl (OH) groups.

Figure 6.21 displays the impact of heating (900°C) on the surfaces of AAMs containing various levels of BGWNP as GBFS replacement. The control sample of AAM prepared with 0% of BGWNP showed number of wider cracks (Figure 6.21a) than those containing 5% and 20% of BGWNP. The deterioration reduced slightly with increasing BGWNP content and showed less surface cracks. Addition of 5% BGWNP to alkali-activated matrix enhanced the specimens' resistance to elevated temperatures and manifested less number of cracks (Figure 6.21b). The specimens prepared with 20% of BGWNP showed less deterioration compared to control sample (Figure 6.21c). In fact, heating process is normally accompanied with several transformations, moisture evaporation, internal vapour pressure, fine aggregate expansion, alkali-activated paste contraction, and chemical decompositions. During early stages of heating, these transformations may not have sufficient ability to cause any crack, except in situations where the heating rate is quite high and that the concrete is either dense or contains enough moisture content in which case, spalling may occur within the first 30 minutes of exposure.

Figure 6.22 presents the effect of BGWNP as GBFS replacement on XRD patterns of AAMs exposed to 900°C. Appearance of semi-crystalline aluminosilicate

FIGURE 6.21   Surface texture of AAMs exposed to 900°C.

gel and quartz (Q) was observed in the samples exposed to elevated temperatures. The broad peaks of all alkali-activated mortars can be seen in the region 25°–30° 2θ. Zeolites formed as secondary reaction product and formed as crystalline phase after fire resistance test was completed. The strong peaks in BGWNP base alkali activated mainly identified the presence of quartz, mullite, and nepheline at 900°C temperature. Mullite was the only stable crystalline phase of Al_2O_3–SiO_2 system. Mullite retained its room temperature strength at elevated temperatures and showed high temperature stability with low thermal expansion and oxidation resistance. After exposure to 900°C, peaks of quartz were still stable wherein mullite peaks become progressively stronger. The phase transition from goethite to hematite has occurred at about 400°C. At this temperature, most of the constitutional water molecules were released. The outgoing flux of OH- groups and the simultaneous diffusive rearrangement of the grain structure may cause a local accumulation of internal stress and may even be capable of fracturing the hematite grains. At this

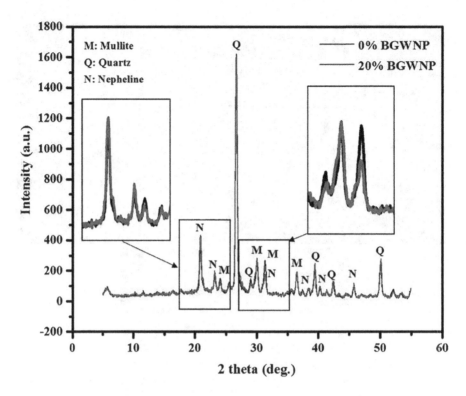

FIGURE 6.22 XRD patterns of AAMs exposed to 900°C containing BGWNP as GBFS replacement.

temperature, grain shape and size of the newly formed hematite phase were still substantially the same as those of original goethite. The XRD pattern of specimen subjected to 900°C disclosed hematite and started to disappear while crystalline nepheline $AlNaSiO_4$ (sodium aluminium silicate) showed its presence in the specimen. Quartz remains the major phase together with mullite phase. In AAM containing 20% of BGWNP, the peaks were stable at elevated temperatures compared to control sample (0% BGWNP).

Figure 6.22 presented the SEM morphology of 0% and 20% BGWNP-contained AAMs which were exposed to 900°C. The porous structures, visible network of micro-cracks, and large pores of AAMs without BGWNP were transformed to the more dense structure with less pores which appeared more durable with the increase in BGWNP content as GBFS replacement. Only few micro-cracks were observed on the surface of specimen containing 20% BGWNP at elevated temperature exposure. The decrease in the CS of AAMs with increasing temperatures could be attributed to the effect of pore structure and the increase in internal cracks which were influenced by heating and cooling processes and induced micro-cracks as evidenced by the reduction of UPV values [37]. Furthermore, as the temperatures increased to 900°C, the dense structure transformed to the progressively less compact structure with a visible network of micro-cracks and large pores (Figure 6.23).

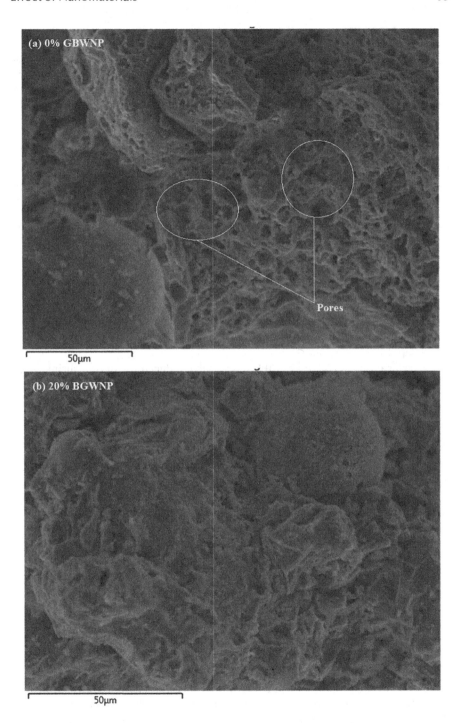

FIGURE 6.23 SEM images showing the effect of elevated temperatures on morphology of AAMs containing BGWNP (a) 0% and (b) 20%.

6.9 SUMMARY

The influenced AAM durability performance in terms of WA, drying shrinkage, carbonation depth, abrasion resistance, resistance to freezing-thawing cycles, sulphuric acid resistance, and resistance to elevated temperatures were evaluated in this chapter. Based on the findings, the following conclusions could be provided:

i. Inclusion of BGWNP within 10% attributed to reduce the porosity and WA which enhanced the mortars' durability. Increase in BGWNP contents to 5% and 20% has enhanced the reaction process and produced denser C–S–H and C–A–S–H gels which improved further the microstructures of mortars with reduced voids.

ii. An inverse relationship was established among the CS and porosity for the AAM specimens. Inclusion of 5% BGWNP as GBFS replacement enhanced the microstructures and CS of the prepared mortars and reduced the porosity by 14.7% compared to the control sample.

iii. Addition of nanomaterials (5%) in AAMs resulted in enhancing the durability performance against the reduction in carbonation depth. A linear relationship was found between the porosity and the carbonation depth.

iv. Inclusion of BGWNP as GBFS replacement reduced the drying shrinkage and enhanced the durability performance.

v. Proposed mortar prepared with 5% of BGWNP displayed high performance related to freezing-thawing cycles. In fact, the reduction in porosity of AAMs increased the resistance to freezing-thawing cycles.

vi. Direct relationship was observed between mortars' abrasion resistance (wearing) and CS. The highest wearing was achieved in specimens containing 5% BGWNP as GBFS replacement.

vii. Proposed mortars containing BGWNP showed quite high resistance to sulphuric acid attack. The specimens' resistance to acid attack increased with increasing BGWNP, where a minimum strength loss of 7.3% was achieved for AAMs containing 20% of BGWNP in the blend.

viii. The detailed microstructures analyses (using XRD) of BGWNP incorporated mortars displayed the formation of less amount of free Portlandite and gypsum, reduction in porosity, and improvement in the durability performance against sulphuric acid attack.

ix. Increase in the BGWNP content enhanced the resistance of AAMs at elevated temperatures in terms of residual strength and weight loss.

x. Results of XRD and SEM explained the thermal stability of AAMs containing high levels of BGWNP (20%) when exposed to heat.

xi. A practical advantage of studying surface discolouration of fired AAMs is the preliminary assessment of damage caused by fire hazards, so that intensity of fire can be comprehended.

REFERENCES

1. Mhaya, A.M., et al., Performance evaluation of modified rubberized concrete exposed to aggressive environments. *Materials*, 2021. **14**(8): p. 1900.
2. Mohammadhosseini, H., et al., Enhanced performance for aggressive environments of green concrete composites reinforced with waste carpet fibers and palm oil fuel ash. *Journal of Cleaner Production*, 2018. **185**: pp. 252–265.
3. Huseien, G.F., et al., Durability performance of modified concrete incorporating fly ash and effective microorganism. *Construction and Building Materials*, 2020: p. 120947.
4. Huseien, G.F., K.W. Shah, and A.R.M. Sam, Sustainability of nanomaterials based self-healing concrete: an all-inclusive insight. *Journal of Building Engineering*, 2019. **23**: pp. 155–171.
5. Guterrez, K., How nanotechnology can change the concrete world, 2005: pp. 1–10.
6. Abdoli, H., et al., Effect of high energy ball milling on compressibility of nanostructured composite powder. *Powder Metallurgy*, 2011. **54**(1): pp. 24–29.
7. Jankowska, E. and W. Zatorski. Emission of nanosize particles in the process of nanoclay blending. in *Quantum, Nano and Micro Technologies, 2009. ICQNM'09. Third International Conference on*, 2009, IEEE.
8. Sanchez, F. and K. Sobolev, Nanotechnology in concrete–a review. *Construction and Building Materials*, 2010. **24**(11): pp. 2060–2071.
9. Shah, S.P., et al., Nanoscale modification of cementitious materials, in Bittnar Z., Bartos P.J.M., Němeček J., Šmilauer V., Zeman J. (eds) *Nanotechnology in Construction 3*. 2009, Springer: Berlin, Heidelberg, pp. 125–130.
10. Saleh, N.J., R.I. Ibrahim, and A.D. Salman, Characterization of nano-silica prepared from local silica sand and its application in cement mortar using optimization technique. *Advanced Powder Technology*, 2015. **26**(4): pp. 1123–1133.
11. Thomas, J.J., H.M. Jennings, and J.J. Chen, Influence of nucleation seeding on the hydration mechanisms of tricalcium silicate and cement. *The Journal of Physical Chemistry C*, 2009. **113**(11): pp. 4327–4334.
12. Thokchom, S., P. Ghosh, and S. Ghosh, Effect of water absorption, porosity and sorptivity on durability of geopolymer mortars. *ARPN Journal of Engineering and Applied Sciences*, 2009. **4**(7): pp. 28–32.
13. Huseien, G.F., et al., Utilizing spend garnets as sand replacement in alkali-activated mortars containing fly ash and GBFS. *Construction and Building Materials*, 2019. **225**: pp. 132–145.
14. Saha, A.K., Effect of class F fly ash on the durability properties of concrete. *Sustainable Environment Research*, 2018. **28**(1): pp. 25–31.
15. Lim, N.H.A.S., et al., Microstructure and strength properties of mortar containing waste ceramic nanoparticles. *Arabian Journal for Science and Engineering*, 2018. **43**(-10): pp. 5305–5313.
16. Samadi, M., et al., Enhanced performance of nano-palm oil ash-based green mortar against sulphate environment. *Journal of Building Engineering*, 2020. **32**: p. 101640.
17. Tanzadeh, J., Laboratory evaluation of self-compacting fiber-reinforced concrete modified with hybrid of nanomaterials. *Construction and Building Materials*, 2020. **232**: p. 117211.
18. Basheer, P., *40 Designs of Concrete to Resist Carbonation*, Vol. 1. 1999, NRC Research Press Ottawa: Canada.
19. Farhana, Z., et al., The relationship between water absorption and porosity for geopolymer paste. *Materials Science Forum*, 2015. **803**: pp. 166–172, Trans Tech Publications Ltd.
20. Huseien, G.F. and K.W. Shah, Durability and life cycle evaluation of self-compacting concrete containing fly ash as GBFS replacement with alkali activation. *Construction and Building Materials*, 2020. **235**: p. 117458.

21. Liu, Y.-W., T. Yen, and T.-H. Hsu, Abrasion erosion of concrete by water-borne sand. *Cement and Concrete research*, 2006. **36**(10): pp. 1814–1820.

22. Wang, S.-D., K.L. Scrivener, and P. Pratt, Factors affecting the strength of alkali-activated slag. *Cement and Concrete Research*, 1994. **24**(6): pp. 1033–1043.

23. Marchand, J., et al., Influence of chloride solution concentration on deicer salt scaling deterioration of concrete. *Materials Journal*, 1999. **96**(4): pp. 429–435.

24. Liu, Z. and W. Hansen, Freezing characteristics of air-entrained concrete in the presence of deicing salt. *Cement and Concrete Research*, 2015. **74**: pp. 10–18.

25. Wu, H., et al., Experimental investigation on freeze–thaw durability of Portland cement pervious concrete (PCPC). *Construction and Building Materials*, 2016. **117**: pp. 63–71.

26. Mayercsik, N.P., M. Vandamme, and K.E. Kurtis, Assessing the efficiency of entrained air voids for freeze-thaw durability through modeling. *Cement and Concrete Research*, 2016. **88**: pp. 43–59.

27. Yusuf, M.O., Performance of slag blended alkaline activated palm oil fuel ash mortar in sulfate environments. *Construction and Building Materials*, 2015. **98**: pp. 417–424.

28. Bhutta, M.A.R., et al., Sulphate resistance of geopolymer concrete prepared from blended waste fuel ash. *Journal of Materials in Civil Engineering*, 2014. **26**(11): p. 04014080.

29. Donatello, S., A. Fernández-Jimenez, and A. Palomo, Very high volume fly ash cements. Early age hydration study using Na_2SO_4 as an activator. *Journal of the American Ceramic Society*, 2013. **96**(3): pp. 900–906.

30. Huseien, G.F., et al., Evaluation of alkali-activated mortars containing high volume waste ceramic powder and fly ash replacing GBFS. *Construction and Building Materials*, 2019. **210**: pp. 78–92.

31. Shah, K.W. and G.F. Huseien, Bond strength performance of ceramic, fly ash and GBFS ternary wastes combined alkali-activated mortars exposed to aggressive environments. *Construction and Building Materials*, 2020. **251**: p. 119088.

32. Ahmed, M.A.B., et al., Performance of high strength POFA concrete in acidic environment. *Concrete Research Letters*, 2010. **1**(1): pp. 14–18.

33. Ariffin, M., et al., Sulfuric acid resistance of blended ash geopolymer concrete. *Construction and Building Materials*, 2013. **43**: pp. 80–86.

34. Bamaga, S., et al., Evaluation of sulfate resistance of mortar containing palm oil fuel ash from different sources. *Arabian Journal for Science and Engineering*, 2013. **38**(9): pp. 2293–2301.

35. Noruzman, A., et al., Strength and durability characteristics of polymer modified concrete incorporating vinyl acetate effluent. *Advanced Materials Research*, 2013. **690**: pp. 1053–1056.

36. Abdulkareem, O.A., et al., Effects of elevated temperatures on the thermal behavior and mechanical performance of fly ash geopolymer paste, mortar and lightweight concrete. *Construction and Building Materials*, 2014. **50**: pp. 377–387.

37. Roviello, G., et al., Fire resistant melamine based organic-geopolymer hybrid composites. *Cement and Concrete Composites*, 2015. **59**: pp. 89–99.

7 Nanoparticle-Based Phase Change Materials for Sustained Thermal Energy Storage in Concrete

7.1 INTRODUCTION

In recent times, constant increase in the fossil fuel prices and emission levels of green-house gases enforced us to use alternative renewable energy sources. The demand on energy increased rapidly around the world and especially in developed countries. Presently, it is very important to seek alternative effective methods to reduce energy usage and dependency on fossil fuels and indirectly switching parts of heavy demands in power consumptions, this will not only prevent more greenhouse gas emissions but will also conserve energy for better utilization. Thus, in the engineering applications, storage of thermal energy using phase change materials (PCMs) has become the priority in build-ing energy conservation [1]. Thermal energy conservation involves the storing of heat collected during daytime that can be utilized in the night-time within the solar energy (photovoltaic) systems to apply in the buildings. By storing the heat, it not only increases the efficiency of power generation system via load levelling but also enhances further the energy conservation at reduced generation cost. In recent development, one of the major approaches in the thermal energy storage (TES) systems is the implementation of PCMs.

As aforementioned, PCMs are by far the most advanced efficient latent heat storage (LHS) system which possesses the characteristics of high TES density and the isother-mal characteristic of heat energy storing procedure. With this property, PCMs have been utilized for heat pumps, photovoltaics, and heat control appliances in the spacecraft [2]. In the past decades, numerous research studies have been performed to exploit PCMs for heating and cooling applications [3]. Of late, various PCMs which could melt and solidify at different temperatures and their versatility have become useful in a number of applications [4–6]. The US Department of Energy's (DOE) Building Technologies Office has set a momentum of producing high-performance, energy-efficient buildings that require cost-effective, durable, energy-efficient building envelopes. Recently, many buildings have embarked on utilizing PCMs due to their high latent heat (LH) capacities in reducing envelope-generated heating and cooling loads.

Zhou et al. [7] reviewed previous research studies on applications of PCMs in build-ings for TES. They studied the characteristics of PCM in depth and its usage in building

DOI: 10.1201/9781003196143-7

construction for energy cost savings and how to provide absolute thermal comfort due to increase in temperature. Various approaches to PCM applications in building envelopes have been investigated: PCM wallboards [8–11], PCM mixed in construction materials like concrete (Figure 7.1) and bricks [12,13], PCM mixed with fibrous insulation in walls and attics [3,14,15], and macro-packaged PCM in plastic pouches [16].

With developments in nanotechnology, the nanoparticles have been used for the enhancement of TES [17,18]. Using LHS in the buildings can meet the demand for thermal comfort and energy conservation purpose. The quests for proper thermal storage materials have lately been targeted to utilize nanomaterials to overcome various issues associated with inorganic PCMs, such as supercooling and phase separation. Considering the enormous potential of NE-PCMs, this chapter assesses the latent TES aspects of NE-PCMs in concrete applications to achieve enhanced performance.

7.2 OVERVIEW OF ENERGY STORAGE

Worldwide, coordinated sun-powered radiation is considered to be the most significant and imminent wellsprings of energy. With regard to this, many researchers around the world are seeking solutions for new sustainable energy resources. One of the possibilities is to develop effective energy storage device(s) in various structures that can be converted into required shape for utilization. Energy storage lessens not only the confusion amongst free market activity but also increases the execution and dependence on conventional energy systems with assured energy conservation [7,19,20]. This prompts sparing of best fills and creates the framework practical by reducing energy wastage, thus making it cost-effective. For instance, energy storage can increase the power generation efficiency of the system by stack levelling, thereby improving the effectiveness via immediate energy security at a lower price. Therefore, implementation of PCMs has

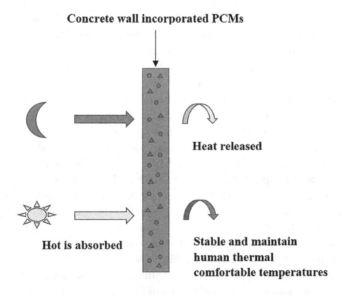

FIGURE 7.1   Warming and cooling capacity of concrete wall consolidated with PCM.

been considered as the beneficial approach for TES. However, before the large-scale usages of such innovative technology, it is essential to evaluate various issues related to this innovation for future sustainable energy security and environmental remedy.

7.2.1 Energy Storage Methods

Currently, several methods are developed for energy storage including mechanical energy storage systems, electrical storage, thermo-chemical energy storage, and TES. The TES systems are accomplished through varied innovations. They allow not only loads of TES but can also be used over a period of hours, days, or months later, either ranging from individual usage to buildings, multiuser buildings in district, town, or region. This usage balanced of energy consumption remains between daytime and night-time, or even storing during summer heat which could be used for winter heating or even winter cold for summer air conditioning (seasonal TES).

Diverse wellsprings of thermal energy for storage include heat or cold produced with heat pumps from off-peak, low-cost electric power, a practice called peak shaving; heat from combined heat and power (CHP) plants; heat produced by renewable electrical energy that exceeds grid demand and waste heat from industrial processes, etc. Heat storage is viewed as an imperative means for efficiently adjusting high offers of variable inexhaustible power creation and reconciliation of power and warming areas in vitality frameworks nearly or totally sustained by sustainable power source. Thermal energy can be stored as a change in internal energy of a material as sensible heat, LH, and thermo-chemical or combination of these. Figure 7.2 illustrates the overview of solar thermal storage methods.

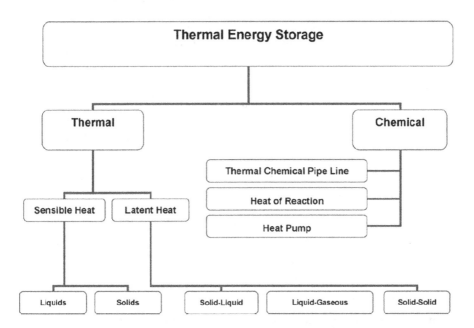

FIGURE 7.2 Various approaches for TES [19].

7.2.2 Latent Heat Storage (LHS)

The LHS of materials depends on the warmth assimilation or discharge when they experience a transformation from solid to fluid or liquid to gas or in reverse. This perspective has been widely recognized and utilized because PCMs can store energy of high density and at fixed temperature in relation to its phase conversion temperature. Phase change conditions can be varied in various structures such as solid to solid, solid to fluid, solid to gas, liquid to gas, and vice versa. In solid to solid conversion, the storage of heat happens when the material begins to crystalline continuously. In comparison, solid to liquid transitions will produce higher LH and higher volume transformation. However, solid to solid phase conversion of PCMs has the advantages of less requirement of strict holder with more prominent plan adaptability. In other structures, solid to gas and liquid to gas transition has the disadvantages of containment issues due to their large volume transformations. In addition, huge transformations in volume cause the system to be more complex and impractical. Furthermore, solid to liquid transformations possess lesser LH than liquid to gas, but these only dictate a minor change (of order of 10% or less) in volume. Among all these changes, solid to liquid transformation has been categorized as most feasible and cost efficient in the TES systems.

In the process of heat transfer, PCMs need a heat exchanger to function as channel to transport energy from source to PCMs and then from PCMs to loads. Heat exchanger must possess low thermal diffusivity of PCMs. This is due to PCMs' volume during dissolving process which requires volume plan to contain the whole of PCM. It should possess the ability to contain this volume transformation easily available for the PCMs to be utilized. Therefore, all the LHS system must have the criteria including (i) appropriate PCMs with desired range of melting temperature; (ii) proper surface for heat exchange, and (iii) right container compatible to the PCMs. In short, the growth of a potential TES system must display three crucial attributes such as right PCMs, materials for containers, and efficient heat exchangers.

7.2.3 Methods for Latent Heat of Fusion (LHF) and Melting Temperature Measurement

Generally, the melting temperature and latent heat of fusion (LHF) of PCMs are determined via (i) differential thermal analyser (DTA), and (ii) differential scanning calorimeter (DSC) [21,22]. To achieve these results, the samples and the appropriate reference materials are heated at constant rates whereby the difference in temperature between DSC/DTA corresponds with the contrasting heat flow among the two materials and the DTA/DSC curves are recorded. In such measurement, alumina (Al_2O_3) is chosen as the reference material. LHF is calculated from the DTA curve and the melting temperature is assessed using the tangent at the point of maximum slope around the peak.

7.2.4 TES System

In the past, TES was applied in various systems such as solar water and air heater, cookers, green house, and buildings. Since 1980s, with the invent of PCMs as TES in buildings, they have been used in tromped walls, wall- and ceiling boards, shutters, under-floor heating systems, and warming and cooling applications. Constant development on PCM wallboard and PCM solid frameworks are occurring to improve the TES capability of usual gypsum wallboards and the concretes, especially during peak load shift and solar energy use. On the same boundary, the main objectives of PCMs usage in construction sector are: firstly, the utilization of solar energy for heating during night cold or cooling during hot temperature at daytime, and secondly, utilizing manmade heat or cold sources. Overall, PCMs are essential as heat and cold storage components in the buildings including walls and other units.

7.3   PHASE CHANGE MATERIALS

A PCM is a substance with a high heat in combination of melting and solidifying at a specific temperature which is capable of storing and releasing large amounts of energy. Heat is absorbed or released when the material changes from solid to liquid and the other way round; accordingly, PCMs are classified as LHS units. As a rule, LHS can be achieved by transforming phases from liquid to solid, solid to liquid, solid to gas, and liquid to gas. However, the suitability during solid to liquid or reverse phase transformation is considered here. Even though liquid to gas transformation produces higher heat than solid–liquid phase changes, it is unsuitable for TES because of their high volume or high pressure in the gas phase. In addition, solid to solid phase transformation is not suitable for PCM materials since it has slow rate of transformation and releases low heat during the process.

Initially, solid to fluid PCMs were identified as suitable stockpiling (SHS) materials, but it was observed that their temperature arises as they retain warm. For customary SHS materials, when PCMs achieve the phase transformation temperature (liquefying temperature), the temperature remains relatively consistent. The PCM keeps on absorbing heat without raising the temperature till the point when all the material is changed to fluid stage. However, when the temperature of fluid drops, the PCMs solidify and release stored LH. PCMs are known to be available over a wide range of temperatures (−5°C to 190°C) [23]. For our ultimate thermal comfort, the desirable PCM temperature is from 20°C to 30°C. Besides, PCM can accumulate 5–14 times higher amount of LH/m³ compared to usual substances such as water, masonry, or rock.

7.3.1   PHASE CHANGE MATERIALS CLASSIFICATION

In the classification of phase changes, numerous PCMs are identified to be utilized at any desired temperature domain. Figure 7.3 presents the classification of PCMs. From the perspective of melting point, several natural and inorganic compounds are recognized as PCMs; however, with the exception of melting point, majority of PCMs do not fulfil the criteria required for a sufficient storage media. Since there is no single material that fulfilled the required properties for an ideal thermal-storage media, one needs to utilize the accessible materials and attempt to compensate for the poor physical property by a sufficient framework plan. For instance, metallic fins can be utilized to increase the thermal conductivity of PCMs. Whereas in the process

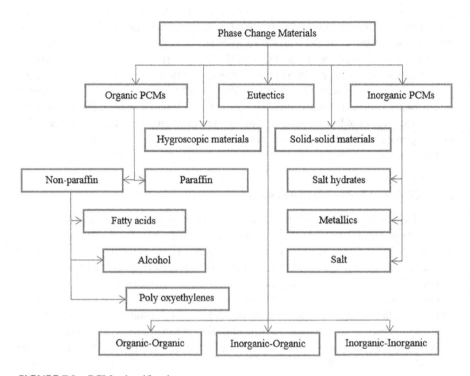

FIGURE 7.3   PCMs classification.

of suppressing the supercooling effect, a nucleating agent of suitable thickness or a 'cold finger' can be utilized to impede the incongruent melting in the storage material. In general, the organic compounds (128–200 kg/dm³) possess half the volumetric LHS capacity as compared to the inorganic compounds (250–400 kg/dm³). The classification of each subgroup for thermal and chemical characteristics and the properties using PCMs are depicted hereunder.

7.3.2   ORGANIC MATERIALS

Organic PCMs (OPCMs) are divided into two groups as paraffins and non-paraffins [24]. They undergo consistent melting without phase segregation and result in degradation of their LH of transformation. In addition, self-nucleation means that they crystallize with very little or no supercooling and are typically non-corrosive. In TES materials, one of the commonly used solid to liquid OPCMs is paraffin. Paraffin consists of fatty acids, alcohol, polyoxyethylenes which have very similar properties among them [25]. Paraffin comprises of saturated hydrocarbons (n-alkanes of formula $C_nH_{2n}O_2$) having essentially the same characteristics as the organic materials. The higher the carbon number, the higher the melting point of paraffin. Additionally, the carbon numbers also influence the thermal features (peaks for melting and crystallization temperature) of paraffin. If the number of carbon atoms in paraffin is less than 19 or over 40, it does not display a secondary peak in the DSC curves.

Conversely, if the number of carbon atoms is in the range of 19 and 40, then the paraffin displays a secondary peak alike nonadecane [26].

Fatty acids are the other class of favourable solid to liquid OPCMs [27,28] with a general chemical formula $CH_3(CH_2)2nCOOH$. The thermal characteristics of fatty acids are dictated by the number of carbon atoms present in the structure. For higher number of carbon atoms, the LH of melting and freezing are higher and vice versa, which are analogous to paraffin [27]. The thermal features of fatty acids are decided by their purity. However, direct use of fatty acids as TES is disadvantages because of bad odour that arise from weak thermal stability and the sublimation when heated [29]. Polyoxyethylenes, called PEGs, are made of linear chains of dimethyl ether with hydroxyl terminating groups in the form $HO-CH_2-(CH_2-O-CH_2-)n-CH_2-OH$. It is a useful OPCM because of appropriate phase transformation temperature and high LHS capacity. Another added advantage is that the thermal attributes of PEGs could be adapted by their molecular weight alteration [30]. Additionally, PEG atoms are biodegradable [31,32]. However, the low thermal conductivity and instability of melting for PEGs limit their major applications. On top, PEG molecules do not mix well with the interfacial molecules of adjacent materials because of dual character such as water solvency as well as organic dissolvability [33].

Polyalcohol (polyol) and alcohol derivatives are the other types of promising OPCMs. They reveal solid to solid phase conversion and have varied thermal characteristics depending on their chain structural symmetry. The tetragonal and monoclinic phases of polyalcohol are responsible for low temperature while the face-centred cubic (FCC) phase leads to high temperature transformations. The FCC phase of polyalcohol can absorb more energy from hydrogen bonding due to higher symmetry than other crystalline structures [34]. Some of the advantages and disadvantages of OPCMs are listed hereunder.

    I. Advantages
        1. Availability in a large temperature range.
        2. High LHF.
        3. No supercooling.
        4. Chemically stable and recyclable.
        5. Good compatibility with other materials
    II. Disadvantages
        1. Low thermal conductivity.
        2. Relative large volume change.
        3. Flammability.

7.3.3 INORGANIC MATERIALS

Inorganic PCMs (IOPCMs) are further classified as salt hydrates and metallics. They do not supercool appreciably and the LHF remains non-degradable through cycling. Salt hydrates (M_nH_2O, where M is as inorganic compound) work as TES system because of their high-energy storing capacity. Although the use of salt hydrates as IOPCM was first reported in 1947 [35], several limitations related to phase conversion behaviours limited their applications as TES systems. Metals are the other potential IOPCMs which overcome some of the problems faced with salt hydrates. For instance, poor thermal conductivity, strong corrosion level, and large variation in the volume during melting can be surmounted using metals [36,37]. Nonetheless, metals are appropriate for high temperature applications above 4100°C. The freezing temperatures of IOPCMs are not well documented in the existing literatures. IOPCMs have the following merits and demerits.

    I. Advantages
       1. High heat of fusion.
       2. High thermal conductivity.
       3. Low volume change.
       4. Availability in low cost.
   II. Disadvantages
       1. Supercooling.
       2. Corrosion.

7.3.4 EUTECTICS MATERIALS

A eutectic is a base dissolving synthesis of at least two components, each of which melts and freezes congruently forming a mixture of crystals during crystallization [38]. This means, eutectic material normally melts and freezes without segregation as they freeze to form a firm mixture of crystals, unable to separate even when both components liquefy simultaneously. Eutectic materials have the following merits and demerits.

    I. Advantages
       1. Sharp melting temperature.
       2. High volumetric thermal storage density
   II. Disadvantages
       1. Lack of currently available test data on thermo-physical properties.

7.3.5 HYGROSCOPIC MATERIALS

Several natural building materials can retain and discharge water easily due to their hygroscopic nature via the following processes:

- Condensation: Gas to liquid phase transformation with change in enthalpy (ΔH) negative which reduces by giving off the heat (warms up) so called exothermic process.
- Vaporization: Liquid to gas phase transformation with ΔH positive which increases by absorbing the heat (cools down) so called endothermic process.

Although these processes liberate a tiny amount of energy, the enlarged surface areas permit substantial (1°C–2°C) warming or cooling in the buildings through insulating material's finishing (wools, earths, or clays).

7.3.6 SOLID TO SOLID PHASE TRANSFORMING MATERIALS

Special class of PCMs transform their crystal structure at a particular temperature, which may involve LH similar to the most efficient solid to liquid phase transforming systems. These materials are valuable, because unlike solid to liquid transforming materials they never undergo nucleation to avoid supercooling. Furthermore, on the grounds that it is a strong solid to solid phase-transforming entity, no noticeable alteration (except a minor extension/compression) in the presence of PCM is found. Besides, no issues related to liquid handling including containment and possible spillage are reported. Currently, the temperatures of solid to solid phase converting PCMs range from 25°C to 180°C.

7.4 CRITERIA OF PCMs TO BE USED FOR TES

One of the main criteria of PCM to be used for TES is that it has the "Latent" warmth stockpiling ability. This happens when materials transform from solid to fluid, or fluid to solid, known as *state* or *phase* transformation. The initial state occurs when this solid to fluid phase changes temperature of PCMs by absorbing heat. In contrast to traditional capacity materials, PCM absorbs and releases heat at a steady temperature and stores 5–14 times higher warmth for each unit volume compared to conventional capacity materials, such as water, brick work, or shake. Numerous PCMs are identified to be able to transform at any required temperature range, and also to function as stockpiling materials, they must possess thermodynamic, kinetic, and suitable chemical properties. In addition, PCMs must also be cost effective and easily assessable. Table 7.1 enlists the criteria of PCM selection for TES application.

7.5 CRITERIA OF TES BASED ON PCMs FOR BUILDING APPLICATIONS

Another vital parameter when selecting PCMs for building applications is the scope of working temperature. The scope of working temperature for selected PCMs must be as per neighbourhood atmosphere, area in the structures, or the specified framework where the PCMs are utilized. It has been identified that PCMs that possess

TABLE 7.1
Criteria to Select PCMs for TES Applications

Thermal properties	Suitable phase-transition temperature, high latent heat of transition, and good heat transfer.
Physical properties	Favourable phase equilibrium, high density, small volume change, and low vapour pressure.
Kinetic properties	No supercooling, sufficient crystallization rate.
Chemical properties	Long-term chemical stability, compatibility with materials of construction, no toxicity, and no fire hazard.
Economy	Abundant, accessible, and cost-effective.

phase transformation of temperature to suit human comfort (18°C–30°C) could be utilized in the buildings [39,40]. Three temperature regions of PCMs were recommended for the building structures including (i) cooling application: zero up to 21°C, (ii) human comfort suitability: 22°C–28°C, and (iii) boiling water applications: 29°C–60°C [41]. It was reported that [42] PCMs with liquefying temperature around the coveted inward diurnal air and temperature of building are reasonable for the ground usage empowering the full warm cycle of PCMs amidst fluid and solid states.

7.6 BENEFITS OF TES BASED ON PCMs FOR BUILDINGS

Utilization of PCMs in the TES framework brings substantial advantages in building construction. This innovation is the most effective approach to store warm vitality in the building development [43]. Some of the fundamental benefits of TES based on PCMs for building constructions are as below.

  i. Enhancing energy use productivity and administration.
  ii. Energy storage for future utilization.
  iii. Warming and cooling load transition.
  iv. Resilient.
  v. It permits a superior incorporation of sustainable power sources framework, for example, sun-based cell frameworks into the electrical frameworks.
  vi. Combination with protection materials.

7.7 TECHNOLOGY, DEVELOPMENT, AND ENCAPSULATION

Amongst all, hydrated salts, fatty acids, esters, and paraffins are the most commonly used PCMs. Lately, ionic liquids have emerged as novel PCMs. Because the greater part of organic solutions is devoid of water, they are effective in air. However, every salt-based PCM requires encapsulation to avoid from evaporating or water uptaking. These factors offer some advantages and drawbacks in the event of accurate connection and often turn into favourable condition for specific purposes. These PCMs have been used since the late 19th century as a media for the TES appliances including refrigerated transport in the railways and roads. Unlike ice storage system, PCMs can be utilized with any regular water chilling applications. The positive phase transforming temperature allows radiation and retention of chillers together with the ordinary respond and screw chiller frameworks or even lesser encompassing situations using a waterless cooler to charge TES structure.

Respecting the intermediate and high-temperature TES applications, the temperatures offered by the PCM innovation provide another skyline to the building administrations and refrigeration engineers. TES system has wide range of applications such as solar heating, water heating, and heating rejection (cooling system and dry cooling-based TES appliances). Because PCMs convert the phases among solid to fluid in the thermal cycle, encapsulation obviously appears the option for storing (Figure 7.4). PCM encapsulations are of various types such as:

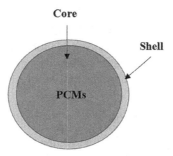

FIGURE 7.4 Encapsulated PCM.

i. Macroencapsulation: Such growth of PCMs with large volume control failed because of low thermal conductivity, where they had solidification tendency at the containers' edges that prevented the efficient thermal transport.

ii. Microencapsulation: Being economic and simple, it permits the PCM's incorporation in construction materials including concrete and provides a portable TES system. Via microscopic PCM coating with the protective layer, the particles are suspended in a continuous phase like water, a system called phase change slurry (PCS).

iii. Molecular encapsulation: This technique was introduced by DuPont de Nemours which permits the incorporation of very high contents of PCM inside polymers to achieve a TES as much as 515 kJ/m^2 (103 MJ/m^3) for a 5 mm board. Furthermore, such encapsulated PCMs can be drilled and cut through without leaking any PCM.

Generally, PCM containers are isolated into cells because of their excellent performance in small storage compartment. These cells must be thin enough to reduce the static head, a rule called shallow geometry of compartment so the substance can conduct the heat well. On top, they must be sufficiently durable to resist frequent variation in the specimens' volume during phase transformation. It must limit the water flow through the walls to keep the THS materials wet in case it is hygroscopic in nature. The leaking and corrosion must be avoided during encapsulation. At room temperature, conventional packing specimens including stainless steel, polypropylene, and polyolefin reveal display of chemical compatibility with PCMs.

7.8 NANOMATERIAL-BASED PCMs

Presently, nanomaterials have been emerged as novel PCMs for TES applications owing to their many unique properties including excellent chemical stability, non-toxic, and suitable range of melting temperatures useful for solar passive heating.

7.8.1 PREPARATION METHODS

Numerous methods have been developed for the preparation of PCM microcapsule including chemical, physical, and physico-chemical processing [44]. Inorganic polymers are generally utilized as shell [45] in the particulate size range of 5–400 µm. The molecular sizes of nanoPCMs are below MicroPCMs. Thus, the conventional synthesis methods of MicroPCMs are unsuitable for nanoPCMs. Presently, the main techniques for synthesis of nanocapsule are based on polymerization including interfacial, emulsion, mini-emulsion, in situ as well as sol–gel processing.

7.8.2 Interfacial Polymerization

For PCMs synthesis via interfacial polymerization (IP) technique, the core material is firstly emulsified followed by the creation of oil/water or water/oil emulsion. Next, the polymer capsules are shaped on the core surface via monomer polymerization. Eventually, capsules are detached from oil or water phases. For nanoPCM synthesis, cores are mixed in the syringe using capillary action at high voltage in the presence of direct current. Additionally, the partition among the syringe needle and fluid level of monomer arrangement should be closer. IP technique is appropriate for the nanoPCM using water as well as oil-soluble PCM core. So far, the nanoPCM created by IP technique utilizes paraffin as core and polyurea as shell.

Park et al. [46] prepared nanoPCM using IP technique with core as paraffin and shell as polyurea. DSC analysis of nanoPCM revealed a melting temperature of 56.54°C, LH melting was 101.1 J/g, freezing point was 47.82°C, and LH freezing was 105.6 J/g. Figure 7.5 displays the SEM and TEM images of the produced nanoPCM. Clearly, the nanoPCM showed spherical morphology with diameter ranging between 400 and 600 nm.

7.8.3 Emulsion Polymerization Method (EP)

Emulsion polymerization is one of the regular strategies for making natural nanoPCM. Here the scattered insoluble monomer is dissolved consistently inside the response media (emulsifier) because of the surfactant and mechanical blending. At that point, the polymer layer is created on the core surface via the inclusion of initiator to start the polymerization in which nanoPCM ultimately appears. EP strategy is frequently used for polymers to be ready for nanoPCM utilizing fluid PCM as core

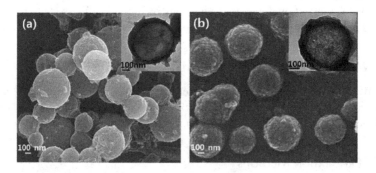

FIGURE 7.5 SEM and TEM micrographs of nanoPCM (a) with and (b) without containing Fe_3O_4 nanoparticles [46].

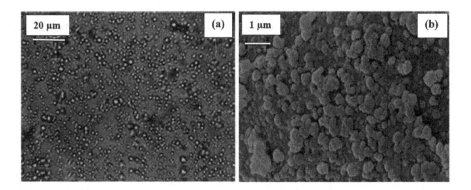

FIGURE 7.6   As-synthesized PS/heptadecane micro/nanoPCM imaged using (a) POM and (b) SEM [47].

matter. Up until this point, nanoPCM has been readied by means of emulsion polymerization technique which usually utilizes alkane for core and polystyrene or poly (methyl methacrylate) for shell. Sari et al. [47] used EP to prepare micro/nanoPCMs in encapsulation with the n-heptadecane and the polystyrene as core and shell materials (at the ratio of 1:2), respectively. DSC curve revealed that the melting temperature, LH melting (LHM), freezing temperature, and LHF of such micro/nanoPCM was 21.371°C, 136.89 J/g 21.481°C, and 134.67 J/g, respectively. LHM was reduced from 136.89 to 128.27 J/g because of numerous capsule damage beyond 5000 thermal cycles. Thermogravimetric analysis (TGA) of these micro/nanoPCM displayed excellent thermal stability for readiness procedure.

Figure 7.6a and b, respectively, shows the polarized optical microscope (POM) and SEM images of prepared micro/nanoPCM [47] where the structure is deficient of spherical morphology with particle size distribution (PSD) ranged between 10 and 40 μm. Later, EP technique was utilized to prepare micro/nanoPCM with n-nonadecane as core and poly (methyl methacrylate) (PMMA) as shell [26]. Again, the PSD ranged between 10 and 40 μm with melting temperature and LHM (obtained from DSC measurements) as 31.231° even beyond 5000 thermal cycles. It was concluded that synthesis of micro/nanoPCM as TES system using PS/n-heptadecane and PMMA/n-nonadecane could be promising for the development of solar thermal controller for building envelopes, thermoregulation-based textiles, thermal protection of vehicles battery, various heat regulation purposes, etc. Back et al. [48] and Alkan et al. [49] also prepared nanoPCM using EP strategy and characterization.

7.8.4   Miniemulsion Polymerization (MEP) Method

This technique is greatly recognized strategy for preparing nanoPCM wherein polymerization reaction is completed inside steady and decentralized small beads estimated at nanoscale under the influence of large shear limit. It contains water, monomer, emulsifier, initiator, etc. During MEP reaction, the monomer decides the chemical constituents of the product and latex features [50]. The generated PE morphologies are alike the original droplets [51,52]. Compared to EP, the MEP technique is advantageous in terms of nanoPCM readiness where less energy is required to achieve high solidity. This strategy is reasonable to combine alkane (PCM core) and polystyrene, polyurea, styrene-butyl acrylate (SBA), styrene (St)-methylmethacrylate (MMA) copolymer, and poly (methyl methacrylate) as shell.

Following MEP technique, Chen et al. [53] prepared nanoPCM containing n-dodecanol and SBA copolymer as core and shell, respectively. The morphology (shape and size) and thermal properties of these nanoPCM were determined using PSD, TEM, and DSC measurements, respectively. Results showed 98.4% exemplification productivity of nanoPCM with circular morphology (Figure 7.7). When the proportion of monomers/n-dodecanol arrived at 1:1, the mean particle diameter was 100 nm with thermal decomposition temperature of 195.1°C, phase transition temperature of 18.41°C, and phase change enthalpy of 109.2 J/g.

Fang et al. [54] used MEP method to prepared nanoPCM with polystyrene shell and n-tetradecane, n-octadecane [55,56], or n-dotriacontane [57] core. Spherical particle morphology was achieved with z-mean particle sizes of 132, 124, and 168.2 nm, respectively. The melting temperature and LHM obtained from DSC analyses were 4.041°C and 98.71 J/g, respectively. The freezing temperature and LHF were 3.431°C and 91.27 J/g, respectively. The LH of n-octadecane/polystyrene nanocapsules was as high as 124.4 J/g whereas the melting temperature, LH solidifying temperature, and inert warmth of n-dotriacontane/polystyrene nanocapsules were 70.91°C and 174.8 J/g, 63.31°C, and 177.1 J/g, respectively.

Using MEP technique, Tumirah et al. [58] prepared nanoPCM containing noctadecane and styrene (St)-MMA copolymer as core and shell, respectively. The

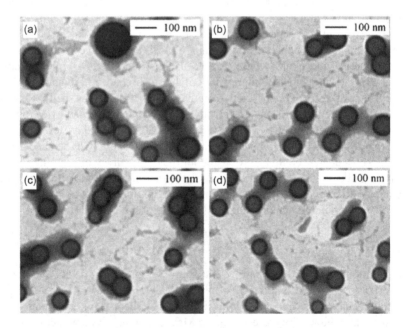

FIGURE 7.7 TEM images of SBA/n-dodecanol nanoPCM prepared using varied emulsifier LAS contents of: (a) 2%, (b) 3%,(c) 4%, and (d) 5% [53].

morphology and thermal characteristics were determined. At shell to core mass ratio of 3:1, the average particle size, melting temperature, and freezing temperature of synthesized nanoPCM were 102 nm, 29.51°C and 24.61°C, respectively. The LHM and LHF were 107.9 and 104.9 J/g, respectively. The synthesized nanoPCM maintained excellent chemical and thermal stability even beyond 360 heating/cooling cycles. Furthermore, MEP was used by Luo et al. [59], Fuensanta et al. [60], Chen et al. [61], and Zhang et al. [62] to prepare nanoPCM of mean size less than 200 nm.

7.8.5 IN SITU POLYMERIZATION (ISP) METHOD

The in situ polymerization (ISP) strategy was used to prepare phase change capsules where the reacting monomer and catalyst were kept exterior of the core. In the continuous phase, the monomer is soluble but the polymer is insoluble, thus the polymerization reaction takes place at the core surface. With the development of polymerization, the pre-polymer is formed slowly on the core surface and eventually the capsule shell is created [63]. So far, using ISP scheme, common polymers including urea-formaldehyde resin, melamine-formaldehyde, carboxymethyl cellulose (CMC), poly (methyl methacrylate), and poly (allyl methacrylate) are largely coated as shell substance.

Hu et al. [64] used ISP to prepare spherical nanoPCM with paraffin (core) and CMC-modified MF (shell). These nanocapsules had mean size of 50 nm. The PC enthalpy increased from 41.79 to 83.46 J/g due to the increase in paraffin mass content from 31.6% to 63.1%, respectively. The corresponding cracking ratio of nanocapsules was reduced from 17.5% to 11.0% when the nanocapsule suspension was subjected to mechanical shear at revolution of 5000 rpm for 10 minutes. Wang et al. [65] prepared the nanoPCM using free-radical EP and ISP technique with poly (methyl methacrylate-co-allyl methacrylate) and n-octadecane as shell and core, respectively.

Figure 7.8 illustrates the SEM images of nanoPCM containing different quantities of polyaniline (PANI). Effects of varied PANI contents (nucleation agent) on the surface morphology, crystallization behaviour, and thermal stability of nanoPCM were examined. SEM analyses showed spherical morphology of capsules with PSD ranging from 100 to 1000 nm and mean diameter between 577 and 693 nm. Inclusion of PANI at lower contents had insignificant effects on the capsules morphology and encapsulation efficiency. However, at higher amount of PANI contents, the degree of supercooling was substantially reduced. The supercooling level of the capsule was varied from 2.31°C to 0.51°C for the PANI and concentration ranged from 0 to 2.0 g, respectively. NanoPCM composed of n-tetradecane (core), and polymerization product of urea and formaldehyde (shell) was synthesized using ISP technique [66]. SEM analysis revealed an average size of 100 nm, where a phase change enthalpy as high as 134.16 J/g was achieved when the mass content of n-tetradecane surpassed 60%.

7.8.6 SOL–GEL METHOD

Sol–gel procedure needs relatively mild preparation condition with the following main processes. First, metal alkoxide as precursor is uniformly mixed with the solvent, catalyst, and complexing agent. Second, a stable and transparent colloid

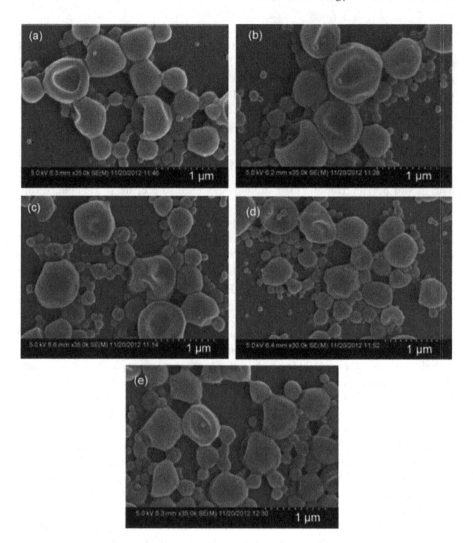

FIGURE 7.8 SEM images of nanoPCM at different PANI content of (a) 0 g, (b) 0.5 g, (c) 1.0 g, (d) 1.5 g, and (e) 2.0 g [65].

mixture appeared after hydrolysis and condensation reaction. Third, the gel with 3D network structure is generated after sol being aged. Ultimately, macro/nanoPCM is extracted by drying, sintering, and curing [67–69]. This procedure is appropriate for nanoPCM preparation containing alkane, palmitic acid (PA), and indium (core) and silicon dioxide (shell).

Using sol–gel procedure, Latibari et al. [70] prepared nanoPCM containing PA as core and SiO_2 as shell at varied pH contents (11, 11.5, and 12). Figure 7.9 displays the SEM images of synthesized nanoPCM with spherical morphology. Fourier transform infrared spectroscopy (FTIR), X-ray diffraction (XRD), and energy dispersive spectroscopy (EDS) analyses revealed their compact and smooth surface. The average

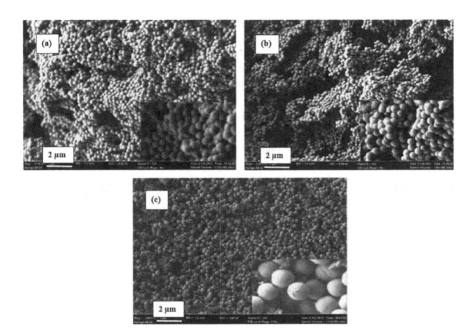

FIGURE 7.9 SEM images of PA/SiO$_2$ nanoPCM prepared at varied solvent pH of: (a) 11, (b) 11.5, and (c) 12 [70].

diameters of nanoPCM obtained from SEM and TEM analyses varied from 183.7 to 722.5 nm when the encapsulation ratios of PA increased from 82.53% to 88.32%, respectively. Besides, the thermal conductivity of nanoPCM was higher than pure PA. A test of 2500 thermal cycling of the optimum nanoPCM sample revealed the variation of melting and freezing temperatures from 61.61°C to 60.11°C and 57.081°C to 56.851°C, respectively. Both the LHM and LHF were reduced from 180.91 to 177.3 J/g and 181.22 to 178.6 J/g, respectively. It was demonstrated that the studied nanoPCM had excellent thermal and chemical stability together with high reliability.

Hong et al. [71] prepared nanoPCM via sol–gel route with silica (shell) and indium (core). Tetraethylorthosilicate (TEOS) and sodium silicate were used to get two kinds of silica, which were further included in nanoPCM. The particle size and the degree of supercooling of these two types of nanoPCM were determined. The research results showed that using TEOS-derived silica as shell in nanoPCM produced its core diameter and shell thickness of 200 and 100 nm, respectively. For other types of silica-based nanoPCM, the corresponding values were 200 mm and 50 nm. The supercooling temperatures for TEOS-derived silica as shell in nanoPCM enhanced from 32.1°C to 41.1°C when the heating rates increased from 1.1 to 40.1°C/ min. For the same heating rates, the supercooling temperatures of the other types of nanoPCM ranged from 3.91°C to 14.1°C.

Amongst all the aforesaid techniques, IP is the most common one for encapsulation of dye and pesticides. This technique is simple with many commercial interests. Nonetheless, IP technique is limited because of proper shell material selection to synthesize nanoPCM. Conversely, nanoPCM obtained via ISP shows excellent particle morphology and thermal traits. More research studies are needed to make this method simpler and cost-effective for large-scale industrial-level production. Yet, MEP strategy is robust to synthesize core/shell polymers [72] where several researchers have successfully used it to prepare nanoPCM with excellent morphology and thermal stability. Following moderate polymerization rate in MEP technique, nanoPCM can easily be attained with higher stability [73], where capsule morphology can be tuned by adjusting the stabilizing doses. This method is greatly beneficial to meet the large-scale industrial demand of nanoPCM.

7.9 PCM INCORPORATION PROCEDURES IN CONCRETE

Few methods have been introduced to include PCM in concrete such as immersion [12,74], impregnation [75,76], and direct mixing [77]. Vacuum impregnation of PCM is performed using porous mass. In the concrete mixing phase, direct piling of encapsulating PCM can be performed. Past studies revealed that due to the incorporation of PCM in concrete, the warmth and mechanical attributes of the product were significantly improved. Following sections describe each of these three techniques briefly.

7.9.1 IMMERSION TECHNIQUE

This procedure was introduced by Hawes [78] wherein the dousing of the permeable concrete creations was included in the molten (liquid) PCM and was termed as immersion PCM-concrete. It was led by inundating permeable concrete product in the liquid PCM-filled container. The viability and the time of submersion of fluid PCM that was splashed inside the permeable cement followed three criteria such as the retention limit of concrete, the temperature, and kinds of PCM. Generally, the submersion procedure took few hours. The effects of various kinds of permeable concrete on the absorption properties of PCM incorporated concrete were examined [12]. It was shown that at $(80 \pm 5)°C$, the inundation time was sufficient to permit the fluid PCM to douse inside the concrete pores. Concrete blocks such as autoclaved and regular were conceptualized. For autoclaved concrete blocks, the time taken for complete dousing of butyl stearate (BS) and PAR ranged from 40 minutes to 1 hour. Customized concrete slabs containing PAR took around 6 hours. This clearly showed that autoclaved concrete slabs were superior for inundation than normal one due to their higher porosity and greater assimilation. Furthermore, the inundation speed was faster when the temperature of the used fluid PCM was higher.

Using DSC measurement, Lee et al. [79] demonstrated great circulation of BS and PAR in autoclaved and customary concrete slabs during drenching. When portion of PAR spilled out between the warming cycles, it produced a thin film of PAR that stayed on the outer surface of the normal concrete slab. Nonetheless, it showed insignificant influence on the synthetic response of the concrete products due to the soundness of solid compound. For real application, extraordinary medicines amongst the submersion of PCM-solid items are fundamental with a specific end goal to keep any dissolved PCM from streaming out concrete. Otherwise, it could create environmental pollution. Salyer [80] recommended the addition of silica particles for stopping the concretes PCM spillage.

7.9.2 IMPREGNATION TECHNIQUE

This strategy includes three straightforward advances [77]. Firstly, air and water are emptied from the permeable or lightweight totals using vacuum pump. Secondly, the permeable totals absorb the fluid PCM inside a controlled domain (in the presence of vacuum). Thirdly, the pre-doused PCM permeable total worked as a transporter for the PCM blended concrete. Zhang et al. [76] assessed PCM absorption capacity of three kinds of porous collections such as expanded clay (C1), normal clay (C2), and expanded shale (S) as the carrier for BS-based PCM. Table 7.2 enlists the essential features and the after effects of PCM retention limit with respect to the studied aggregates. A general pattern of expanding PCM assimilation limit was observed when the thickness of the porous aggregate was reduced, implying high porosity. Regarding the water absorption capacity, the vacuum impregnation strategy was better than the basic inundation system. Furthermore, the distinction was more articulated in the C2 than in S. The estimated PCM absorption ability of C1, C2, and S aggregates was 0.876, 0.176, and 0.081 mL/g, respectively. These results indicated that PCM could possess up to 75% of the aggregate pore space for C1.

Using DSC, Bentz and Turpin [77] examined the thermal flow of lightweight aggregate soaked in PEG and PAR-based PCM. The maximum temperature of PAR at liquefying and cementing was marginally greater than PEG. Besides, absorbing permeable totals PCM could upgrade the warmth exchange among the PCM in porous and bulk concrete structures.

7.9.3 DIRECT MIXING TECHNIQUE

In this technique, before the PCM is consolidated [81] into concrete, it must be embodied inside a synthetically and physically stable shell. This holds the PCM in its unadulterated shape and guarantees the absence of impedance with the solid constitutes. Likewise, when the PCM is in a fluid state (in dissolution), the containers ensure absence of any fluid PCM spilling. The well-recognized strategies for

---

TABLE 7.2
Fundamental Characteristics of Various Porous Aggregates and Their PCM-Absorption Capacity [76]

Aggregates	Density (g/m³)	Porosity (%)	Water Absorption Ability for Simple Immersion (%)	Water Absorption Ability for Vacuum Impregnation (%)	PCM Absorption Ability of Porous Aggregate (mL/g)
C1	0.76	75.6	11.0	72.5	0.876
C2	1.25	41.9	5.9	42.5	0.176
S	1.39	33.8	4.1	15.0	0.081

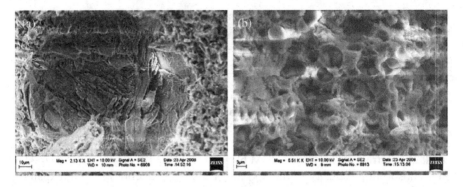

FIGURE 7.10 (a) SEM images of open pores enclosed with solid wax and (b) a magnified (5500×) view showing a matrix portion with distorted and ruptured microcapsules containing partly leaked pure wax [17,81].

typifying the natural PCM are the IP [82], EP [83], ISP [84], and additionally spray drying and co-acervation [85].

For coordinate blending, the surface (shell) hardness of the PCM microcapsules should be indestructible and feasible to stay away from any breakage (harm), amidst the concrete blending and casting procedure. To keep any untimely burst in these phases, layer fortification items, for example, zeolite, or zeocarbon (for the most part got from charcoal and volcanic shake) can be utilized for surface support to resist high contact or effect [85]. As of late, an endeavour was made to evaluate the achievability of utilizing distinctive measures of Micronal DS 5008 X (smaller scale embodied paraffin wax PCM) as the immediate swap for marble powder in the generation of self-compacting concrete. It was shown that the permeable small-scale structures and circular voids of the cracked PCM microcapsules could be seen using SEM imaging. Figure 7.10 displays the SEM images where the chipped structures of the set wax were found to secure the inward mass of the pore. A close examination of the pore divider revealed that an extensive volume of cell dividers had smooth earmarks [81].

7.10   EFFECTS OF CPM ON CONCRETE PROPERTIES

7.10.1   FRESH PROPERTIES

Hunger et al. [81] analysed the impact of direct mixing of microencapsulated PCMs (MCPCMs) on the fresh characteristics of self-consolidated concrete (SCC) using the slump flow, V-funnel time, and J-ring tests. It was concluded that better self-compacting properties of concrete could be achieved using PCM microcapsules. The stream distances across all the PCM– SCC blends were comparable with plain SCC blend under varying contents of water and superplasticizer. It was shown that the SCC blends with 3% and 5% of PCM microcapsules had marginally higher thickness, which was majorly ascribed to the higher water content, and sizes of PCM microcapsules that affected the workability of solid blend [86]. Bentz and Turpin [77] examined the temperature increase and decrease of bond mortars under semi-adiabatic situations. Three kinds of mortars were considered such as a control with cement to nonporous coarse sand and water to cement ratio of 3 and 0.40, respectively. A mix containing pre-impregnated lightweight aggregate with 100% PAR as PCM and another having pure PAR particles of 100% as total substitute of aggregate on a volumetric basis was chosen. It was found that the normal for warm development of PCM mortar was unequivocally affected due to the nearness of PCM. Utilization of PCM showed a decrease in the most extreme temperature increase and the lessening was more articulated when the PCM was specifically included to the blend. Likewise, utilizing paraffin wax PCM in the bond mortar resulted in 1-hour delay in achieving the pinnacle temperature, meaning the hindrance of concrete hydration.

Recently, extended clay and porous Greek pumice aggregates were combined [87] with PAR, PEG, and vegetable oil-based PCM into cement mortars. It was observed that PAR had insignificant impact on hydration response, while the PEG and vegetable oil-based PCM revealed significant impact on impeding (delay in achieving peak temperature) and smothering (lowering in crest temperatures) the hydration of the concrete grid. For all PEG blends, it was discovered that not just the most extreme warmth stream diminished by around 40%, but also the warmth development had been additionally deferred by a few hours. Furthermore, it was proposed that the measures of non-retained PCM following on the totals surface were probably going to be an issue amidst bond hydration. Comparative outcomes for warm advancement on account of direct blending were likewise detailed by Hunger et al. [81].

7.10.2 MECHANICAL CHARACTERISTICS

The mechanical properties, especially the compressive strength of PCM concrete, have been widely investigated. It was shown that the PCM content and the methods for PCM joining in concrete recognizably influence the compressive quality of the PCM concrete. Examinations on utilizing the drenching system have discovered that there is no huge distinction in the quality amidst control and inundation of PCM concrete. Hawes [78] revealed that when PAR is in the fluid state, the PCM concrete displayed a comparable quality with control concrete. Despite the control, concrete had half addition in quality even when PAR was in the strong state. Cabeza et al. [13] showed an enhancement in PCM concrete and accomplished a compressive strength of more than 25 MPa. A tractable part quality of more than 6 MPa (at 28 days) that had fitting levels for some auxiliary application purposes.

Regarding direct blending, consideration of PCM microcapsules in SCC essentially weakens the compressive strength [81] which was reduced by 13% for each extra level of MCPCM inclusion to concrete weight. This reduction in strength could be attributed to the following (i) major difference between the inherent strength of the microcapsules and other concrete specimens (e.g. cement paste) and (ii) damage of MCPCM due to paraffin wax leakage from the ruptured microcapsules and blending with other concrete specimens. Hunger et al. [81] examined the concrete porosity (as part of the void fraction which is open to the surface) of a control SCC and PCM–SCC blends containing 1%, 3%, and 5% of MCPCM and added to cement weight. They revealed that the PCM–SCC with MCPCM possessed higher porosity than the control SCC. This may be because of the auxiliary difference in the solid pressing thickness. The natural solid thickness of the PCM–SCC diminished with an expansion in the MCPCM content, most likely because of the generally low specific gravity of the MCPCM (0.915 kg/m^3) when contrasted with different specimens in concrete.

7.10.3 DURABILITY PROPERTIES

Wei et al. [88] assessed the impact of PCM on the durability performance of concrete. This investigation examined the effect of microencapsulated PCM's dose on the toughness of cementitious materials. It is noticed that, PCMs encountered a huge enthalpy diminishment (25%) when subjected to sulfate-bearing conditions while staying impassive in antacid arrangements. The response to sulfate particles showed shell crack after paraffinic PCM core was in contact to sulfate ions, causing enthalpy reduction. With respect to water sorption, PCMs fill in as non-sorptive consideration like reviewed quartz sand, hence expanding the volume either by considering a decrease in volume or water absorption rate. These impacts, particularly the degree of terminal water sorption, can be evaluated in case of non-sorptive incorporations utilizing Powers' model [89].

The dry shrinkage of cementitious concrete remained the same in presence of PCMs. Conversely, addition of stiff quartz lowered the shrinkage significantly because of aggregates' moderation. Hobbs model explained both influences appropriately such as inclusion of stiffness and volume fraction which provided a way to determine the shrinkage of cementitious composites with such constituents. Generally, when PCMs are influenced remarkably in sulphate-based cementitious atmosphere, their durability remains unaffected.

7.11 STABILITY OF PCM IN CONCRETE

The alkalinity level of cement is an important parameter that must be contemplated while consolidating PCM in concrete. It was realized that not all PCMs were appropriate for use in high-soluble base-level cement in light of the fact that the antacid can debase some PCMs. Hawes [78] explored the alkalinity of different solid composites by deciding the pH estimations of water wherein the concrete samples were submerged for 28 days. The outcomes showed that the autoclaved concrete slabs possessed minimum alkali concentration, followed by other samples with extended shale aggregates (EXSLs). It was discerned that butyl stearate (BS), and dodecanol (DD) were more appropriate for concrete to lower alkalinity levels. Besides, PRA was very steady in all the concrete specimens because it was synthetically latent in the soluble media. This was estimated by the DSC strategy and affirmed by materials' investigation. The most pessimistic scenario was to utilize polyethylene glycol (PEG) PCM which had a tendency to fall apart in most concretes.

Some types of concrete such as pumice concrete slab seem to have a high-assimilation limit and are perfect for PCM addition where the solid needs to decrease the soluble base substance in these concrete specimens. Hawes [78] proposed to utilize pozzolans to decrease the alkalinity of cement. Nonetheless, the amounts of pozzolans' utilization must be controlled at a suitable level to guarantee that the PCM holds adequate porosity. Hawes [78] examined the freeze-thawing strength of inundation PCM concrete using different solid slabs and PCMs. The test was carried out at an interval of 6 hours for each cycle at the temperature range of 33°C to −25°C for a total of 50 cycles. Insignificant dissimilarity was seen in the freeze-thawing cycles of the autoclaved specimens (ABL) containing various types of PCM (BS, PAR, and PEG). It was noticed that the PCM concrete would perform better to protect it from freeze-thawing cycles compared to the control samples without PCM because of less water absorption and lower penetrability.

Lee [74] reported that the weight loss of autoclaved samples containing PAR after 20 freeze-thawing cycles was much below the general samples incorporating PAR. In case of autoclaved samples, the PCM weight loss was around 5%, while for general samples, it was around 31%. This could be due to concretes' dense structure where a large portion of PAR stayed at its surface. Similarly, in the freeze-thawing cycles, PCM weight loss in the normal concrete samples was substantially higher than the autoclaved concrete samples. Bentz and Turpin [77] utilized a computer model to re-enact the quantity of expected stop/defrost cycles to disappointment of a control cement and PCM concrete. The outcomes exhibited that the nearness of PCM in concrete diminished the quantity of cycles by around 30% by and large when compared with the control concrete without PCM. PCM has been ended up being the most reasonable and successful LHS material for use in concrete. In any case, one of the downsides of PCM is that it might be combustible (e.g. paraffin).

Hawes [78] examined the imperviousness of PCM concrete when subjected to fire. For this reason, distinctive kinds of PCM solid with dimensions of $200 \times 200 \times 30$ mm were tried at 700°C fire for 10 minutes. In this fire test, it was discovered that there was a lessening of PCM mass and increment of the danger of smoke release for the PCM concrete. The shade of the PCM solid surface was likewise influenced by the fire. The rate of PCM mass loss for the regular concrete blocks was lower than the autoclaved one. This observation was attributed to the higher surface tension for smaller pore diameter in regular concrete blocks than the autoclaved one. The measure of unmistakable smoke released went from immaterial to direct contingent upon the sort of PCM utilized as a part of the concrete. Concrete products can end up combustible if the PCM utilized is combustible, isn't embodied, and is specifically blended in concrete in high fixations. Full-scale embodiment or miniaturized-scale exemplification could be conceivable answers for increment of the PCM-concrete protection from flame. BASF has exhibited that PAR PCM fused with a magnesium oxide-based grid can enhance its protection from flares and fire. These materials could pass the Euro-class B fire rating.

7.12 EFFECTS OF NANOMATERIALS ON ENHANCEMENT TES OF PCM

Various approaches were adopted to enhance the thermal performance of PCM such as brushes, fines, metal foam, and fine particles. Nanotechnology is one of the latest methods applied to enhance the CPM performance. Due to nanotechnology growth, the size of the additives [90] is diminished to the nanoscale and further lowering in size can upgrade the suspension execution, particular surfaces, and warmth exchange execution of the added substances. In past investigations, carbon nanotubes (CNTs)

[90–92], carbon nanofibres (CNFs) [92,93], Al_2O_3 nanoparticles [94–96], and Ag nanoparticles [97] were included to prepare nanocomposite-enhanced phase-change materials (NEPCMs) as a method to improve the thermal traits of PCMs. However, CNTs, CNFs, and Ag remain expensive.

Li Min [98] reported that the heat transmission of low-cost paraffin enhanced with nano-graphite (NG) is useful for energy storage technology. Multi-walled CNTs (MWCNTs) dispersed in 1-dodeconal with a concentration of 0, 1, and 2 wt% were also studied [99]. Although heat conduction was intensified due to dispersion of nanoparticles, the viscosity increased under the effect of decreasing buoyancy-driven natural convection. Melting was shown to be decelerated for two different loading weight concentrations of nanoparticles as the degradation in natural convection (resulting from viscosity augmentation) exceeded the enhancement in conduction (due to increase in thermal conductivity). Furthermore, the experimental results were in contrast to the documented numerical results where an acceleration of melting was achieved.

Attempts were made to improve the thermal conductivity and shape stability of paraffin by adding exfoliated graphite nanoplatelets (GnP) [100]. At 10% of GnP, thermal conductivity increased by more than ten times. The low cost of x GnP and high thermal conductivity make it appropriate selection for PCM applications. The addition of Cu, Al, and C/Cu nanoparticles to the melting paraffin resulted in increased thermal conductivity, with a decrease in the rate of heating and cooling by 30.3% and 28.2% for 1 wt% of Cu/paraffin composite, respectively [101]. Another study showed that the thermal conductivity of PCM and the charging and discharging time of LTES have also been reduced significantly when dispersing CuO nanoparticles within the PCM [102]. Al_2O_3 nanoparticles were considered as nanoparticles embedded in paraffin [94]. It was found that the addition of 5 and 10 wt%, respectively, of Al_2O_3 nanoparticles had a little effect of melting/freezing behavior of PCM. From the effective thermal conductivity point of view, a non-linear behaviour with the weight fraction had been detected, relative promotion was found with temperature rise.

The melting of solid–liquid PCMs (n-octadecane) included in dispersed Al_2O_3 nanoparticles was examined [94]. It was concluded that the natural convection could dominate the heat transfer rate across the melted region, and thus, the total energy transfer tends to degrade greatly with increasing mass fraction of nanoparticles [94]. It was argued that such degradation as the enhancement in the thermal conductivity of the nanoPCM could be outweighed by far greater enhancement in the dynamic viscosity of liquid nanoPCM due to disbursement of Al_2O_3 nanoparticles. Another material employed as an additive is the nanomagnetite (Fe_3O_4) for paraffin composites with enhanced thermal conductivity [103]. Sol–gel method was used for nanomagnetite preparation and mixed with paraffin in 10% and 20% mass fraction. The results showed increase in thermal conductivity by 48% and 67% for 10% and 20% of nanomagnetite for the PNMC, respectively. On the other hand, a slight improvement in LH capacity was noticed while the melting temperature of the composite was stable. After 500 cycles of thermal stability test, there was no change in thermal storage capacity.

Saeed et al. [104] also reported improvement in thermal conductivity of PCM by adding Fe_2O_3 nanoparticles. The nanomagnetite was synthesized using chemical precipitation method with a diameter range 16.6–27.2 nm. The thermal behaviour of paraffin/Fe_3O_4 was measured using DSC for three different ratios of 1, 5, and 10 wt%. The DSC results of LH showed a slight increase in 1 wt% of nanoparticle loading and a decrease in 5, and 10 wt%. LH of paraffin/Fe_3O_4 showed an improvement for 1 wt%, while insignificant effect was detected at 5 wt%. Specific heat capacity for solid and liquid states showed a decrease for 10 wt%, and insignificant variation for the other levels. Minor increase in the activation energy was obtained at specific values of 1 and 5 wt% for solid–solid transformation and 5 wt% for solid–liquid conversion. The reason behind the thermal behaviour changes due to nanoparticles could be explained using their high surface area to the volume ratio and surface defect of the nanomagnetite because of various coordination geometries. However, at specific values of nanoparticles loading, Fe_3O_4 can be regarded as potential candidate for TES applications.

Teng et al. [105] evaluated the nanocomposite-upgraded PCMs (NEPCMs) utilizing the immediate blend technique where paraffin was mixed with alumina (Al_2O_3), titania (TiO_2), silica (SiO_2), and zinc oxide (ZnO) as the exploratory examples. The exploratory outcomes exhibited that the TiO_2 is more compelling than other materials in upgrading both the heat conduction and heat stockpiling execution of paraffin for majority of the test parameters. Different nanofluids can be utilized in traditional heat exchangers as a part of structures. Examination demonstrates that nanofluid applications could bring about volumetric stream diminishment, lessening in the mass flow rate, and pumping power reserve funds. Besides, nanofluids required littler warming frameworks with a specific end goal to be equipped for conveying a similar measure of warm vitality, along these lines lessening the size and the underlying expense of gear. This will decrease the arrival of toxins to the earth because of a diminishment in control utilization and the waste delivered towards the finish of the heat exchange framework lifecycle. In cooling frameworks, nanofluids can be utilized as a part of place of chilled water, which is usually utilized as a part of aerating and cooling channels. However, this application has not been investigated broadly.

7.13   APPLICATIONS OF NE-PCMs IN CONCRETE

Concrete being the most popular material (yearly generation around 11 billion metric tons) received the celebrated quote "Man expends no material with the exception of water in such colossal amounts" [106]. The large warm concrete mass walls may be effective mainly in direct atmospheres where it can be utilized to store vitality amidst the day and discharge it amidst evening time in this manner lessening the requirement for assistant cooling and warming [77]. Besides, the vitality stockpiling limit of cement can additionally be upgraded by mixing PCM in concrete blends. Thermocrete, a PCM improved solid, joins a suitable PCM with a solid framework creating concrete with basic and thermostatic characteristics [107]. Concrete is viewed as appropriate for consolidation of PCM in view of the following [78]:

  i. Most celebrated development materials.
 ii. Can be casted into different shapes and sizes.
iii. Possesses large area and small depth of heat exchange.

   iv. Can trade heat at countenances and core surfaces with any mixes.

   v. Can hold PCM by capillary and surface tension forces.

   vi. Can effectively achieve the production and quality control.

   vii. Ease of testing.

Incorporation of PCMs, specifically in concrete, has demonstrated some encouraging outcomes through lower heat conductivity and expansion in heating the mass at particular temperatures. In any case, PCM concrete has demonstrated some negative properties, such as bring down quality, dubious long haul strength, and lower imperviousness to fire [74]. On the other hand, a few examinations on PCM concrete led to beneficial outcomes through lessening indoor temperatures in hot atmospheres [17]. Merging of concrete structures with PCMs was tried in various ways, wherein holes were drilled in the concrete before being filled with PCM [5]. Royon et al. [92] examined the possibility of filling effectively in empty concrete floor with PCMs. The concrete was filled with paraffin PCM at a temperature of 27.5°C. This test demonstrated that the temperature on the opposite side of the empty concrete zone brought down summer conditions. Subsequently, such floors can be utilized as inactive heat conditioner in the late spring. However, more investigations are needed to prove these effects in genuine atmosphere conditions.

Baetens et al. [108] demonstrated that the PCM-improved cement has generally the heat limit of ten times over gypsum wallboards. A portion of this study was quickly depicted by different analysts. The heat execution of cement mortars arranged with n-octadecane/extended graphite composite PCM was assessed [109]. The level of n-octadecane/extended graphite composite PCM in TES cement mortar (TESCM) shifted from 0.5 to 2.5 wt%. The compressive strength and thermal conductivity dropped with the augmentation in levels of n-octadecane/expanded graphite composite PCM in TESCM. The most extreme reductions in compressive strength and thermal conductivity were 55% and 15.5%, respectively. From thermal performance study on small test room (dimensions $100 \times 100 \times 100$ mm), it was concluded that TESCM containing n-octadecane/extended graphite composite PCM could reduce the indoor temperature significantly. Also, the abatement in the expansion of indoor temperature was comparable with the n-octadecane/extended graphite composite PCM in TESCM. It was inferred that TESCM containing n-octadecane/extended graphite composite PCM could be useful for building applications.

The thermal properties of PCMs in concrete were examined by Hawes et al. [110], wherein BS, dodecanol, paraffin, and tetradecanol were utilized with various concrete blocks such as autoclaved, expanded shale, ordinary Portland cement (OPC), pumice, and regular. The impact of concrete alkalinity, methods for PCM consolidation, PCM types, concrete temperature, inundation time, and PCM weakening on PCM retention among the impregnation procedure were studied. It was demonstrated that heat stockpiling can be expanded up to around 300%, and altering the solid by utilizing pozzolans made it conceivable to utilize high alkaline concrete. Hawes et al. [29] studied the thermal behaviour of concrete slabs fabricated with various organic PCMs (BS, dodecanol, paraffin, and tetradecanol). Coordinate fuse and inundation systems were utilized as methods for PCM consolidation, and relying on the sort of concrete block utilized, up to 20% by weight of PCM was retained. The TES ability of PCM concrete slab was in the range of 200%–230% to that of traditional block through 6°C change. Moreover, in contrast with traditional blocks, PCM concrete blocks indicated enhanced toughness to freezing/thawing cycles, similar flexural strength, decreased water absorption, and high imperviousness with insignificant fire spread.

7.14 CONCRETE THERMAL ENERGY STORAGE WITH NE-PCMs

Over the years, different techniques were developed to improve the thermal performance of PCMs. These include extension of surface, fins and heat pipes, micro and macroencapsulation, and inclusion of nanoparticles of high thermal conductivity in the PCM. These nanoparticles can improve the PCM fluid properties. Similar to PCMs, the NEPCMs can be utilized in the building structures for both active and passive purposes, thereby enhancing the overall thermal mass of the building. Besides, these structures could be effective for heating, cooling, and various other systems. They can reduce the mismatch between energy supply and demand by changing and decreasing the maximum load. Numerous efforts were made to develop suitable NE-PCMs for building functions and remain an open challenge to researchers. Some of the significant experiments are described hereunder.

Constantinescu et al. [111] examined the thermal properties of nanocomposite made of PEGs combined with epoxy resin and aluminium powder. They constructed appropriate nanoPCM composite for building purposes to enhance the thermal behaviour. Analyses of thermodynamic properties of various nano-enhanced PCM were carried out for cooling the buildings [112]. It was acknowledged that by growing the nanoparticle contents in PCM-based fluid, the rate of solidification and melting could be improved. Kumaresan et al. [113] utilized water-based nanofluid PCM (NFPCM) to enhance the cooling efficiency for building.

Parameshwaran et al. [114] studied the thermal behaviour of nanoPCM consisting of silver and titanium nanoparticles and compared the results with pure PCM for the TES use in building. It was shown that the freezing time of nano-enhanced PCM was below the pure one, indicating a thermal conductivity and thermal storage potential increase of nanoPCM. Later, silver nanoPCM was included [115] in building air conditioner to examine the efficiency throughout the year. Results revealed an average energy reduction in on-peak and per day of about 36%–58% and 24%–51%, respectively. Harikrishnan et al. [116] prepared a novel composite PCM by blending lauric acid (LA) and stearic acid (SA) with TiO_2, ZnO, and CuO nanoparticles at varied weight fractions. Sari et al. [117] synthesized micro- and nanocapsules of PMMA/capric–stearic acid eutectic mixture via EP method and achieved LHM of 116.25 J/g. Amin et al. [118] synthesized a nano-enhanced PCM with beeswax as the PCM base fluid and graphene nanoparticles for enhancing the thermal properties of PCM useful for buildings.

7.15 ENVIRONMENTAL EFFECTS

A quick overall economic advancement prompts a rapidly expanding energy system. Traditional fossil fuels are constrained and their usage has been identified with outflow of destructive gases, thereby responsible for atmospheric changes and ecological contamination. For the time being, TES frameworks are basic for decreasing reliance on the petroleum products and after that adding to a more effective ecological agreeable vitality utilize. With ever-growing demands of thermal comfort in buildings, the energy use is also advancing in both the domestic and the commercial sectors. The domestic and industrial building sectors are the leading energy users worldwide and consume 28% of the overall energy [119]. To tackle such challenges, energy assets must be used more competently and wisely. TES expert should either utilize sensible heat stockpiling or potentially the inert heat stockpiling. Sensible heat stockpiling has been utilized for a considerable length of time by developers to store and discharge heat energy inactively, yet a bigger material volume is needed for storing a similar measure of vitality in contrast with the dormant heat stockpiling material.

The standard of utilizing the PCMs is basic as the heat is supplied to the material altering its phase from solid to fluid continuously at steady temperature until the entire solid is converted to fluid. Additionally, when heat is discharged, the material transforms the phase from fluid to solid again at constant temperature till the total fluid convert to solid [10]. Conventional building structures have extensive heat idleness (sensible TES) and regular ventilation in the rooms. The main idea is to diminish the wall thicknesses (the weight and material consumption), the transportation expenditures, and the construction period. The LHS via PCM inclusion into some building structures emerged as a promising method for compensating the low storage capacity for largely modernized buildings. The principle inconvenience of light weight structures is the low heat mass control for the substantial temperature variances because of outside warming or cooling load. Utilizing PCM material in these structures can diminish the temperature change, especially because of sun-based radiations as demonstrated in a few numerical investigations [120,121].

Nowadays, nano-encapsulation innovation is quite favourable in utilizing PCM in building structures. Using this approach, PCM incorporating gypsum board, mortar, concrete, or potentially other wall coverage components can be used as integrated parts of the building structure even for light-weight building architectures towards effective TES. A few types of structures with PCMs have already been produced for active and passive solar applications in buildings for direct heat gain. However, the surfaces of the most commercial materials are deficient in delivering heat to the building when the PCM is melted via direct sun light [13]. Thus, liquid migration limits the PCM applications. Some types of packing have been used to overcome this problem. These capsules comprised of tiny containers that could pack the core material within a hard shell. In short, NE-PCMs reveal several significant benefits wherein the capsules can handle phase conversion materials as core by tolerating the immense volume alteration.

7.16   ENERGY AND SUSTAINABILITY

LH TES utilizes PCMs as storage media, which undergoes phase change when energy is absorbed or released. In the sections described so far, the emphasis of the research reported was on determination of optimum PCM-mortar combination for improved energy efficiency of buildings. But every energy solution has its own cost. This decides its utility though the solution may be technically superior. The reported work has very limited information regarding the economic costs because of its escalation due to the introduction of PCMs in the building mortars. The research indicated that an improved thermal behaviour of building elements through PCM incorporation reduces usage of heating and cooling equipment but comprehensive research has not been reported yet by comparing costs associated with different alternatives at different temperate zones.

Shafie-khah et al. [122] have conducted experimental studies with hybrid PCM in mortar and also conducted cost analysis based on energy savings accrued with the option being tried. It was concluded that this hybrid PCM, when incorporated into walls and other structures of the building, supplemented the effect of household management system in terms of cost for a specific demand response program. Cabeza et al. [123] in their review paper have emphasized that Life Cycle Assessment (LCA), Life Cycle Energy analysis (LCEA), and Life Cycle Cost Analysis (LCCA) could be conducted for buildings to determine their environmental impacts during different phases of life cycle of building. This paper mentions that the buildings constructed in urban areas were only considered for environmental evaluation through LCA, LCEA, and LCCA techniques. It has been acknowledged that buildings with PCM included in their walls were more environment-friendly when compared to walls with no PCM [124,125]. Besides, there exists consistency in the LCA analysis conducted for PCM-mortar-based constructions when the same boundary conditions, building systems, and methodologies were considered.

7.17 SUGGESTIONS FOR FUTURE WORKS

Present review focused on the characterization and the TES enhancement of diverse nanomaterials. Further research studies are needed for better understanding of nanomaterial behaviour. The existing analyses implied that most of the nanomaterials added to PCMs could be effective for TES enhancement of various PCMs. To evaluate the effect of nanomaterials on various properties of concrete and especially for durability, careful studies should be carried out. Future research must find the important parameters that affect the TES of nanomaterials. The TES can be a complex function of particle morphology (shape and size), agglomeration, poly-dispersity, etc. Also rare data is available to explain the effects of nanomaterials of TES on mechanical properties such as compressive strength, flexural, and tensile strengths. Durability and sustainability of concrete containing nanomaterials for enhancing the TES require more experimental investigations and analytical explanations. Achieving the customized nanoparticle-based product is the main challenge. Presently accessible nanoparticles are inadequate in terms of precise specifications with controlled properties. In short, the advancement on the nanoparticles growth technique could be useful for the future PCM research.

7.18 SUMMARY

This chapter emphasized the accessible TES technology with PCMs for sundry applications. These technologies have shown advantages for the future energy saving, sustainable development, and environmental remedy. The past development, ongoing research activities, and future trends involving the impact of nanomaterials on concrete properties were highlighted. Furthermore, various strategies for the production of NE-PCM together with their environmental friendliness were underlined. Based on this comprehensive overview, the following conclusions could be drawn:

i. The conceivable utilization of nanomaterials PCMs for TES in concrete is beneficial. The change of the heat warmth stockpiling of PCM-cement could make it versatile for the development and building applications. However, present PCM concrete has some drawbacks that need to be overcome. Unwanted properties such as poor quality, indeterminate long haul security, and low imperviousness to fire that limit the fitting of PCM writes and methods for consolidation need to be improved.

ii. The consequences of various examinations demonstrate that nanomaterial PCM concrete accomplish compressive quality esteems inside the range suitable for most development purposes. On account of direct blending, the quality and thickness are observably diminished as the substance of PCM in the solid increments. PCM concrete arranged by methods for submersion demonstrated comparative quality to those of the ordinary control blend.

iii. The incorporation of PCM in solid produces a critical change in the heat execution of the solid. It is due to the improvement of the heat protection (bring down heat conductivity) and heat mass. Field examinations revealed additionally that the work space dividers with PCM had a littler temperature extend and consequently enhanced warm latency.

iv. Existing literature studies revealed that the enhanced heat conductivity of nanomaterials is the major driving parameters for effective execution in various applications.

v. Most of the nanomaterials have positive effect on enhancing TES of concrete and reduction in energy demand.

vi. As nanomaterials enhance the performance of PCMs and increase the capacity of TES of concrete that led to reduce of pollution from other resources and affect positively to clean environments.

REFERENCES

1. Chuah, T., D. Rozanna, A. Salmiah, S. Thomas Choong, and M. Sa'ari, Fatty acids used as phase change materials (PCMs) for thermal energy storage in building material applications. Jurutera, July 2006: pp. 8–15.
2. Biswas, K., J. Lu, P. Soroushian, and S. Shrestha, Combined experimental and numerical evaluation of a prototype nano-PCM enhanced wallboard. *Applied Energy*, 2014. **131**: pp. 517–29.
3. Shrestha, S.S., et al., Modeling PCM-enhanced insulation system and benchmarking EnergyPlus against controlled field data. Oak Ridge National Laboratory (ORNL); Building Technologies Research and Integration Center, 2011.
4. Bergles, A.E., Enhanced heat transfer: endless frontier, or mature and routine? *Applied Optical Measurements*, 1999: pp. 3–17, Springer.

5. Sonneveld, P., W. Visscher, and E. Barendrecht, The influence of suspended particles on the mass transfer at a rotating disc electrode. *Non-Conducting Particles. Journal of Applied Electrochemistry*, 1990. **20**: pp. 563–574.

6. Andersen, P., R. Muller, and C. Tobias, The effect of suspended solids on mass transfer to a rotating disk. *Journal of the Electrochemical Society*, 1989. **136**: pp. 390–399.

7. Zhou, D., C.-Y. Zhao, and Y. Tian, Review on thermal energy storage with phase change materials (PCMs) in building applications. *Applied Energy*, 2012. **92**: pp. 593–605.

8. Darkwa, K., P. O'Callaghan, and D. Tetlow, Phase-change drywalls in a passive-solar building. *Applied Energy*, 2006. **83**: pp. 425–35.

9. Zhou, G., Y. Zhang, X. Wang, K. Lin, and W. Xiao, An assessment of mixed type PCM-gypsum and shape-stabilized PCM plates in a building for passive solar heating. *Solar Energy*, 2007. **81**: pp. 1351–1360.

10. Kuznik, F. and J. Virgone, Experimental assessment of a phase change material for wall building use. *Applied energy*, 2009. **86**: pp. 2038–2046.

11. Zhou, G., Y. Yang, X. Wang, and J. Cheng, Thermal characteristics of shape-stabilized phase change material wallboard with periodical outside temperature waves. *Applied Energy*, 2010. **87**: pp. 2666–2672.

12. Hawes, D. and D. Feldman, Absorption of phase change materials in concrete. *Solar Energy Materials and Solar Cells*, 1992. **27**: pp. 91–101.

13. Cabeza, L.F., C. Castellon, M. Nogues, M. Medrano, R. Leppers, and O. Zubillaga, Use of microencapsulated PCM in concrete walls for energy savings. *Energy and Buildings*, 2007. **39**: pp. 113–119.

14. Kosny, J., E. Kossecka, A. Brzezinski, A. Tleoubaev, and D. Yarbrough, Dynamic thermal performance analysis of fiber insulations containing bio-based phase change materials (PCMs). *Energy and Buildings*, 2012. **52**: pp. 122–131.

15. Childs, K.W. and T.K. Stovall, Use of phase change material in a building wall assembly: a case study of technical potential in two climates, 2012.

16. Kośny, J., K. Biswas, W. Miller, and S. Kriner, Field thermal performance of naturally ventilated solar roof with PCM heat sink. *Solar Energy*, 2012. **86**: pp. 2504–2514.

17. Ling, T.-C. and C.-S. Poon, Use of phase change materials for thermal energy storage in concrete: an overview. *Construction and Building Materials*, 2013. **46**: pp. 55–62.

18. Barlak, S., O.N. Sara, A. Karaipekli, and S. Yapıcı, Thermal conductivity and viscosity of nanofluids having nanoencapsulated phase change material. *Nanoscale and Microscale Thermophysical Engineering*, 2016. **20**: pp. 85–96.

19. Sharma, A., V.V. Tyagi, C. Chen, and D. Buddhi, Review on thermal energy storage with phase change materials and applications. *Renewable and Sustainable Energy Reviews*, 2009. **13**: pp. 318–345.

20. Zalba, B., J.M. Marın, L.F. Cabeza, and H. Mehling. Review on thermal energy storage with phase change: materials, heat transfer analysis and applications. *Applied Thermal Engineering*, 2003. **23**: pp. 251–283.

21. Giro-Paloma, J., M. Martínez, L.F. Cabeza, and A.I. Fernández, Types, methods, techniques, and applications for microencapsulated phase change materials (MPCM): a review. *Renewable and Sustainable Energy Reviews*, 2016. **53**: pp. 1059–1075.

22. Li, W., R. Hou, H. Wan, P. Liu, G. He, and F. Qin, A new strategy for enhanced latent heat energy storage with microencapsulated phase change material saturated in metal foam. *Solar Energy Materials and Solar Cells*, 2017. **171**: pp. 197–204.

23. Sarkar, M.N.I., A.I. Sifat, S.S. Reza, and M.S. Sadique, A review of optimum parameter values of a passive solar still and a design for southern Bangladesh. *Renewables: Wind, Water, and Solar*, 2017. **4**: p. 1.

24. Khadiran, T., M.Z. Hussein, Z. Zainal, and R. Rusli, Advanced energy storage materials for building applications and their thermal performance characterization: a review. *Renewable and Sustainable Energy Reviews*, 2016. **57**: pp. 916–928.

25. Huang, L., M. Petermann, and C. Doetsch, Evaluation of paraffin/water emulsion as a phase change slurry for cooling applications. *Energy*, 2009. **34**: pp. 1145–1155.
26. Sarı, A., C. Alkan, A. Biçer, A. Altuntaş, and C. Bilgin. Micro/nanoencapsulated n-nonadecane with poly (methyl methacrylate) shell for thermal energy storage. *Energy Conversion and Management*, 2014. **86**: pp. 614–621.
27. Yuan, Y., N. Zhang, W. Tao, X. Cao, and Y. He, Fatty acids as phase change materials: a review. *Renewable and Sustainable Energy Reviews*, 2014. **29**: pp. 482–498.
28. Rozanna, D., A. Salmiah, T. Chuah, R. Medyan, S. Thomas Choong, and M. Sa ari, A study on thermal characteristics of phase change material (PCM) in gypsum board for building application. *Journal of Oil Palm Research*, 2005. **17**: p. 41.
29. Mei, D., B. Zhang, R. Liu, Y Zhang, and J. Liu, Preparation of capric acid/halloysite nanotube composite as form-stable phase change material for thermal energy storage. *Solar Energy Materials and Solar Cells*, 2011. **95**: pp. 2772–2777.
30. Qian, T., J. Li, H. Ma, and J. Yang, Adjustable thermal property of polyethylene glycol/diatomite shape-stabilized composite phase change material. *Polymer Composites*, 2016. **37**: pp. 854–860.
31. Pielichowski, K. and K. Flejtuch, Differential scanning calorimetry studies on poly (ethylene glycol) with different molecular weights for thermal energy storage materials. *Polymers for Advanced Technologies*, 2002. **13**: pp. 690–696.
32. Qian, Y., P. Wei, P. Jiang, Z. Li, Y. Yan, and J. Liu, Preparation of a novel PEG composite with halogen-free flame retardant supporting matrix for thermal energy storage application. *Applied Energy*, 2013. **106**: pp. 321–327.
33. Tang, B., J. Cui, Y. Wang, C. Jia, and S. Zhang, Facile synthesis and performances of PEG/SiO$_2$ composite form-stable phase change materials. *Solar Energy*, 2013. **97**: pp. 484–492.
34. Sarier, N. and E. Onder, Organic phase change materials and their textile applications: an overview. *Thermochimica Acta*, 2012. **540**: pp. 7–60.
35. Khadiran T., Nano-encapsulated organic phase change material as thermal energy storage medium, 2015.
36. Kenisarin, M.M., High-temperature phase change materials for thermal energy storage. *Renewable and Sustainable Energy Reviews*, 2010. **14**: pp. 955–970.
37. Liu, M., W. Saman, and F. Bruno, Review on storage materials and thermal performance enhancement techniques for high temperature phase change thermal storage systems. *Renewable and Sustainable Energy Reviews*, 2012. **16**: pp. 2118–2132.
38. Tyagi, V.V. and D. Buddhi, PCM thermal storage in buildings: a state of art. *Renewable and Sustainable Energy Reviews*, 2007. **11**: pp. 1146–1166.
39. Tyagi, V., S. Kaushik, S. Tyagi, and T. Akiyama, Development of phase change materials based microencapsulated technology for buildings: a review. *Renewable and Sustainable Energy Reviews*, 2011. **15**: pp. 1373–1391.
40. Ravikumar, M. and P. Srinivasan, Phase change material as a thermal energy storage material for cooling of building. *Journal of Theoretical & Applied Information Technology*, 2008. **4** (6): pp. 1–12.
41. Kalnæs, S.E. and B.P. Jelle, Phase change materials and products for building applications: a state-of-the-art review and future research opportunities. *Energy and Buildings*, 2015. **94**: pp. 150–176.
42. Sage-Lauck, J. and D. Sailor, Evaluation of phase change materials for improving thermal comfort in a super-insulated residential building. *Energy and Buildings*, 2014. **79**: pp. 32–40.
43. Oliver, A., Thermal characterization of gypsum boards with PCM included: thermal energy storage in buildings through latent heat. *Energy and Buildings*, 2012. **48**: pp. 1–7.
44. Haris, P.I. and D. Chapman, *New Biomedical Materials: Basic and Applied Studies*. 1998, IOS Press: Netherlands, pp. 1–22.

45. Pan, W., J. Ye, G. Ning, Y. Lin, and J. Wang, A novel synthesis of micrometer silica hollow sphere. *Materials Research Bulletin*, 2009. **44**: pp. 280–283.

46. Park, S., et al., Magnetic nanoparticle-embedded PCM nanocapsules based on paraffin core and polyurea shell. *Colloids and Surfaces A: Physicochemical and Engineering Aspects*, 2014. **450**: pp. 46–51.

47. Sarı, A., C. Alkan, D.K. Döğüşcü, and A. Biçer, Micro/nano-encapsulated n-heptadecane with polystyrene shell for latent heat thermal energy storage. *Solar Energy Materials and Solar Cells*, 2014. **126**: pp. 42–50.

48. Baek, K.H., J.Y. Lee, and J.H. Kim, Core/shell structured PCM nanocapsules obtained by resin fortified emulsion process. *Journal of Dispersion Science and Technology*, 2007. **28**: pp. 1059–1065.

49. Alkan, C., A. Sarı, and A. Karaipekli, Preparation, thermal properties and thermal reliability of microencapsulated n-eicosane as novel phase change material for thermal energy storage. *Energy Conversion and Management*, 2011. **52**: pp. 687–692.

50. Galindo-Alvarez, J., et al., Miniemulsion polymerization templates: a systematic comparison between low energy emulsification (Near-PIT) and ultrasound emulsification methods. *Colloids and Surfaces A: Physicochemical and Engineering Aspects*, 2011. **374**: pp. 134–141.

51. Wang, H., M. Wang, and X. Ge, Graft copolymers of polyurethane with various vinyl monomers via radiation-induced miniemulsion polymerization: influential factors to grafting efficiency and particle morphology. *Radiation Physics and Chemistry*, 2009. **78**: pp. 112–118.

52. Sajjadi, S. and F. Jahanzad, Comparative study of monomer droplet nucleation in the seeded batch and semibatch miniemulsion polymerisation of styrene. *European Polymer Journal*, 2003. **39**: pp. 785–794.

53. Chen, C., Z. Chen, X. Zeng, X. Fang, and Z. Zhang, Fabrication and characterization of nanocapsules containing n-dodecanol by miniemulsion polymerization using interfacial redox initiation, *Colloid and Polymer Science*. 2012. **290**: pp. 307–314.

54. Fang, Y., H. Yu, W. Wan, X. Gao, and Z. Zhang, Preparation and thermal performance of polystyrene/n-tetradecane composite nanoencapsulated cold energy storage phase change materials. *Energy Conversion and Management*, 2013. **76**: pp. 430–436.

55. Fang, Y., S. Kuang, X. Gao, and Z. Zhang, Preparation and characterization of novel nanoencapsulated phase change materials. *Energy Conversion and Management*, 2008. **49**: pp. 3704–3707.

56. Fang, Y., S. Kuang, X. Gao, and Z. Zhang, Preparation of nanoencapsulated phase change material as latent functionally thermal fluid. *Journal of Physics D: Applied Physics*, 2008. **42**: p. 035407.

57. Fang, Y., X. Liu, X. Liang, H. Liu, X. Gao, and Z. Zhang, Ultrasonic synthesis and characterization of polystyrene/n-dotriacontane composite nanoencapsulated phase change material for thermal energy storage. *Applied Energy*, 2014. **132**: pp. 551–556.

58. Tumirah, K., M. Hussein, Z. Zulkarnain, and R. Rafeadah, Nano-encapsulated organic phase change material based on copolymer nanocomposites for thermal energy storage. *Energy*, 2014. **66**: pp. 881–890.

59. Luo, Y. and X. Zhou, Nanoencapsulation of a hydrophobic compound by a miniemulsion polymerization process. *Journal of Polymer Science Part A: Polymer Chemistry*, 2004. **42**: pp. 2145–2154.

60. Fuensanta, M., U. Paiphansiri, M.D. Romero-Sánchez, C. Guillem, Á.M. López-Buendía, and K. Landfester, Thermal properties of a novel nanoencapsulated phase change material for thermal energy storage. *Thermochimica Acta*, 2013. **565**: pp. 95–101.

61. Chen, Z.-H., F. Yu, X.-R. Zeng, and Z.-G. Zhang, Preparation, characterization and thermal properties of nanocapsules containing phase change material n-dodecanol by miniemulsion polymerization with polymerizable emulsifier. *Applied Energy*, 2012. **91**: pp. 7–12.

62. Zhang, G., S. Bon, and C. Zhao, Synthesis, characterization and thermal properties of novel nanoencapsulated phase change materials for thermal energy storage. *Solar Energy*, 2012. **86**: pp. 1149–1154.
63. Chen, Z. and G. Fang, Preparation and heat transfer characteristics of microencapsulated phase change material slurry: a review. *Renewable and Sustainable Energy Reviews*, 2011. **15**: pp. 4624–4632.
64. Hu, X., Z. Huang, and Y. Zhang, Preparation of CMC-modified melamine resin spherical nano-phase change energy storage materials. *Carbohydrate Polymers*, 2014. **101**: pp. 83–88.
65. Nan, G.-H., J.-P. Wang, Y. Wang, H. Wang, W. Li, and X.X. Zhang, Preparation and properties of nanoencapsulated phase change materials containing polyaniline. *Acta Physico-Chimica Sinica*, 2014. **30**: pp. 338–344.
66. Fang, G., H. Li, F. Yang, X. Liu, and S. Wu, Preparation and characterization of nanoencapsulated n-tetradecane as phase change material for thermal energy storage. *Chemical Engineering Journal*, 2009. **153**: pp. 217–221.
67. Jamekhorshid, A., S. Sadrameli, and M. Farid, A review of microencapsulation methods of phase change materials (PCMs) as a thermal energy storage (TES) medium. *Renewable and Sustainable Energy Reviews*, 2014. **31**: pp. 531–542.
68. Kaneko, R., E. Suzuki, M. Jikei, and M.-A. Kakimoto, Preparation and properties of hyperbranched aromatic polyamide-silica composites by sol-gel method. *High Performance Polymers*, 2002. **14**: pp. 105–114.
69. Macwan, D., P.N. Dave, and S. Chaturvedi, A review on nano-TiO₂ sol–gel type syntheses and its applications. *Journal of Materials Science*, 2011. **46**: pp. 3669–3686.
70. Latibari, S.T., M. Mehrali, M. Mehrali, T.M.I. Mahlia, and H.S.C. Metselaar, Synthesis, characterization and thermal properties of nanoencapsulated phase change materials via sol–gel method. *Energy*, 2013. **61**: pp. 664–672.
71. Hong, Y., et al., Controlling supercooling of encapsulated phase change nanoparticles for enhanced heat transfer. *Chemical Physics Letters*, 2011. **504**: pp. 180–184.
72. Zhang, Y., H. Chen, X. Shu, Q. Zou, and M. Chen, Fabrication and characterization of raspberry-like PSt/SiO₂ composite microspheres via miniemulsion polymerization. *Colloids and Surfaces A: Physicochemical and Engineering Aspects*, 2009. **350**: pp. 26–32.
73. Fonseca, G.E., T.F. McKenna, and M.A. Dube, Miniemulsion vs. conventional emulsion polymerization for pressure-sensitive adhesives production. *Chemical Engineering Science*, 2010. **65**: pp. 2797–2810.
74. Lee, T., Latent and sensible heat storage in concrete blocks, School for Building, Concordia University, 1998.
75. Hadjieva, M., R. Stoykov, and T. Filipova, Composite salt-hydrate concrete system for building energy storage. *Renewable Energy*, 2000. **19**: pp. 111–115.
76. Zhang, D., Z. Li, J. Zhou, and K. Wu, Development of thermal energy storage concrete. *Cement and Concrete Research*, 2004. **34**: pp. 927–934.
77. Bentz, D.P. and R. Turpin, Potential applications of phase change materials in concrete technology. *Cement and Concrete Composites*, 2007. **29**: pp. 527–532.
78. Hawes, D.W., D. Banu, and D. Feldman, Latent heat storage in concrete. *Solar Energy Materials*, 1989. **19**: pp. 335–348.
79. Lee, T., D. Hawes, Dt. Banu, and D. Feldman, Control aspects of latent heat storage and recovery in concrete. *Solar Energy Materials and Solar Cells*, 2000. **62**: pp. 217–237.
80. Salyer, I.O., Dry powder mixes comprising phase change materials. Google Patents, 1993.
81. Hunger, M., A.G. Entrop, I. Mandilaras, H. Brouwers, and M. Founti, The behavior of self-compacting concrete containing micro-encapsulated phase change materials. *Cement and Concrete Composites*, 2009. **31**: pp. 731–743.

82. Cho, J.-S., A. Kwon, and C.-G. Cho, Microencapsulation of octadecane as a phase-change material by interfacial polymerization in an emulsion system. *Colloid and Polymer Science*, 2002. **280**: pp. 260–266.

83. Sarı, A., C. Alkan, A. Karaipekli, and O. Uzun, Microencapsulated n-octacosane as phase change material for thermal energy storage. *Solar Energy*, 2009. **83**: pp. 1757–1763.

84. Song, Q., Y. Li, J. Xing, J. Hu, and Y. Marcus, Thermal stability of composite phase change material microcapsules incorporated with silver nano-particles. *Polymer*, 2007. **48**: pp. 3317–3323.

85. Hawlader, M., M. Uddin, and M.M. Khin, Microencapsulated PCM thermal-energy storage system. *Applied Energy*, 2003. **74**: pp. 195–202.

86. Park, S.-K., J.-H.J. Kim, J.-W. Nam, H.D. Phan, and J.-K. Kim, Development of anti-fungal mortar and concrete using Zeolite and Zeocarbon microcapsules. *Cement and Concrete Composites*, 2009. **31**: pp. 447–453.

87. Sakulich, A.R. and D.P. Bentz, Incorporation of phase change materials in cementitious systems via fine lightweight aggregate. *Construction and Building Materials*, 2012. **35**: pp. 483–490.

88. Wei, Z., et al., The durability of cementitious composites containing microencapsulated phase change materials. *Cement and Concrete Composites*, 2017. **81**: pp. 66–76.

89. Powers, T.C. and T.L. Brownyard, Studies of the physical properties of hardened Portland cement paste. *Journal Proceedings*, 1946: pp. 101–132.

90. Shaikh, S., K. Lafdi, and K. Hallinan, Carbon nanoadditives to enhance latent energy storage of phase change materials. *Journal of Applied Physics*, 2008. **103**: p. 094302.

91. Wang, J., H. Xie, and Z. Xin, Thermal properties of paraffin based composites containing multi-walled carbon nanotubes. *Thermochimica Acta*, 2009. **488**: pp. 39–42.

92. Cui, Y., C. Liu, S. Hu, and X. Yu, The experimental exploration of carbon nanofiber and carbon nanotube additives on thermal behavior of phase change materials. *Solar Energy Materials and Solar Cells*, 2011. **95**: pp. 1208–1212.

93. Elgafy, A. and K. Lafdi, Effect of carbon nanofiber additives on thermal behavior of phase change materials. *Carbon*, 2005. **43**: pp. 3067–3074.

94. Ho, C.J. and J. Gao, Preparation and thermophysical properties of nanoparticle-in-paraffin emulsion as phase change material. *International Communications in Heat and Mass Transfer*, 2009. **36**: pp. 467–470.

95. Teng, T.P., B.G. Lin, and Y.Y. Yeh, Characterization of heat storage by nanocomposite-enhanced phase change materials. *Advanced Materials Research: Trans Tech Publications*, 2011: pp. 1448–1455.

96. Arasu, A.V. and A.S. Mujumdar, Numerical study on melting of paraffin wax with Al_2O_3 in a square enclosure. *International Communications in Heat and Mass Transfer*, 2012. **39**: pp. 8–16.

97. Zeng, J., et al., Study of a PCM based energy storage system containing Ag nanoparticles. *Journal of Thermal Analysis and Calorimetry*, 2007. **87**: pp. 371–375.

98. Li, M., A nano-graphite/paraffin phase change material with high thermal conductivity. *Applied Energy*, 2013. **106**: pp. 25–30.

99. Zeng, Y., L.-W. Fan, Y.-Q. Xiao, Z.-T. Yu, and K.-F. Cen, An experimental investigation of melting of nanoparticle-enhanced phase change materials (NePCMs) in a bottom-heated vertical cylindrical cavity. *International Journal of Heat and Mass Transfer*, 2013. **66**: pp. 111–117.

100. Shi, J.-N., et al., Improving the thermal conductivity and shape-stabilization of phase change materials using nanographite additives. *Carbon*, 2013. **51**: pp. 365–372.

101. Wu, S., D. Zhu, X. Zhang, and J. Huang, Preparation and melting/freezing characteristics of Cu/paraffin nanofluid as phase-change material (PCM). *Energy & Fuels*, 2010. **24**: pp. 1894–1898.

102. Karunamurthy, K., K. Murugumohankumar, and S. Suresh, Use of CuO nano-material for the improvement of thermal conductivity and performance of low temperature energy storage system of solar pond. *Digest Journal of Nanomaterials and Biostructures*, 2012. **7**: pp. 1833–1841.

103. Sahan, N. and H.O. Paksoy, Thermal enhancement of paraffin as a phase change material with nanomagnetite. *Solar Energy Materials and Solar Cells*, 2014. **126**: pp. 56–61.

104. Saeed, F., et al., Nanomagnetite enhanced paraffin for thermal energy storage applications. *Digest Journal of Nanomaterials & Biostructures (DJNB)*, 2017. **12**(2): pp. 273–280.

105. Teng, T.-P. and C.-C. Yu, Characteristics of phase-change materials containing oxide nano-additives for thermal storage. *Nanoscale Research Letters*, 2012. **7**: p. 611.

106. Monteiro, P., *Concrete: Microstructure, Properties, and Materials*. 2006, McGraw-Hill Publishing: Singapore, pp. 1–18.

107. Fenollera, M., J.L. Míguez, I. Goicoechea, J. Lorenzo, and M. Ángel Álvarez, The influence of phase change materials on the properties of self-compacting concrete. *Materials*, 2013. **6**: pp. 3530–3546.

108. Jelle, B.P., The role of accelerated climate ageing of building materials, components and structures in the laboratory. in *Proceedings of the 7th Nordic Conference on Construction Economics and Organisation 2013*, Trondheim, Norway, 12–14 June 2013, pp. 111–122.

109. Zhang, Z., G. Shi, S. Wang, X. Fang, and X. Liu, Thermal energy storage cement mortar containing n-octadecane/expanded graphite composite phase change material. *Renewable Energy*, 2013. **50**: pp. 670–675.

110. Hawes, D., D. Banu, and D. Feldman, Latent heat storage in concrete. II. *Solar Energy Materials*, 1990. **21**: pp. 61–80.

111. Constantinescu, M., et al., Latent heat nano composite building materials. *European Polymer Journal*, 2010. **46**: pp. 2247–2254.

112. Kalaiselvam, S., R. Parameshwaran, and S. Harikrishnan, Analytical and experimental investigations of nanoparticles embedded phase change materials for cooling application in modern buildings. *Renewable Energy*, 2012. **39**: pp. 375–387.

113. Kumaresan, V., P. Chandrasekaran, M. Nanda, A. Maini, and R. Velraj, Role of PCM based nanofluids for energy efficient cool thermal storage system. *International Journal of Refrigeration*, 2013. **36**: pp. 1641–1647.

114. Parameshwaran, R., K. Deepak, R. Saravanan, and S. Kalaiselvam, Preparation, thermal and rheological properties of hybrid nanocomposite phase change material for thermal energy storage. *Applied Energy*, 2014. **115**: pp. 320–330.

115. Parameshwaran, R. and S. Kalaiselvam, Energy conservative air conditioning system using silver nano-based PCM thermal storage for modern buildings. *Energy and Buildings*, 2014. **69**: pp. 202–212.

116. Harikrishnan, S., M. Deenadhayalan, and S. Kalaiselvam, Experimental investigation of solidification and melting characteristics of composite PCMs for building heating application. *Energy Conversion and Management*, 2014. **86**: pp. 864–872.

117. Sarı, A., C. Alkan, and A.N. Özcan, Synthesis and characterization of micro/nano capsules of PMMA/capric–stearic acid eutectic mixture for low temperature-thermal energy storage in buildings. *Energy and Buildings*, 2015. **90**: pp. 106–113.

118. Amin, M., N. Putra, E.A. Kosasih, E. Prawiro, R.A. Luanto, and T. Mahlia, Thermal properties of beeswax/graphene phase change material as energy storage for building applications. *Applied Thermal Engineering*, 2017. **112**: pp. 273–280.

119. Sarlos, G. and A. Dauriat, Energy, a challenge for humanity in the 21st century. in *International Conference on Energy and the Environment*, 2003.

120. Hoes, P. and J. Hensen, The potential of lightweight low-energy houses with hybrid adaptable thermal storage: comparing the performance of promising concepts. *Energy and Buildings*, 2016. **110**: pp. 79–93.
121. Ibanez, M., A. Lázaro, B. Zalba, and L.F. Cabeza, An approach to the simulation of PCMs in building applications using TRNSYS. *Applied Thermal Engineering*, 2005. **25**: pp. 1796–1807.
122. Shafie-Khah, M., et al., Optimal behavior of responsive residential demand considering hybrid phase change materials. *Applied Energy*, 2016. **163**: pp. 81–92.
123. Cabeza, L.F., L. Rincón, V. Vilariño, G. Pérez, and A. Castell, Life cycle assessment (LCA) and life cycle energy analysis (LCEA) of buildings and the building sector: a review. *Renewable and Sustainable Energy Reviews*, 2014. **29**: pp. 394–416.
124. Kylili, A. and P.A. Fokaides, Life cycle assessment (LCA) of phase change materials (PCMs) for building applications: a review. *Journal of Building Engineering*, 2016. **6**: pp. 133–143.
125. Tatsidjodoung, P., N. Le Pierrès, and L. Luo, A review of potential materials for thermal energy storage in building applications. *Renewable and Sustainable Energy Reviews*, 2013. **18**: pp. 327–149.

8 Concrete Coatings
Applications of Nanomaterials and Nanotechnology

8.1 INTRODUCTION

The cement (main constituent of the concrete) making factories being the primary contributors to the change of climate and unsustainable development remain the major global concern. In addition to the United States, the developing countries including China, India, Indonesia, Korea, and Turkey are also responsible for the greenhouse gas emission and global warming. Several studies [1–5] reported that the cement industry all over the world that produce the ordinary Portland cement (OPC) more than 4 billion metrics annually as the concretes constituent are mainly responsible for CO_2 emission. Presently, finding a suitable and practical alternative to the OPC is highly challenging. In the construction applications, the OPC is mainly exploited as effective binder for the concretes and other building materials. Usually, the manufacturing industries of the OPC are the main emitter of the atmospheric greenhouse gases [6–9]. The International Energy Agency (IEA) suggested the limit of total CO_2 emissions up to 6%–7% [9–12]. The global demand of OPC is estimated to rise up to 200% by the end of 2050 [9]. Thus, it is essential to mitigate the CO_2 releases from the OPC-associated activities. To achieve this goal, improve the durability performance and the life service of concretes, produce high-performance coating materials, alternative self-healing, and smart construction materials with sustainability, and environmental-friendly traits are urgently required [13–15].

During the service life of cement-based material, the influences of environment would lead to the decrease of its quality by the migration of harmful substances inwards/outwards of its surface [16–18]. It has been well recognized that the densification of the surface of hardened cement-based material will result in a significant improvement of the quality of the entire structure [19,20]. Nowadays, different preventative measures are claimed to prevent, or at least to reduce, the deterioration and steel corrosion in concrete [21]. In this wide context, concrete surface treatments offer a possible way to improve concrete structures' durability [22,23].

A number of surface treatments have been suggested as a solution for making both new and current concrete structures more durable. Based on EN 1504 [4], these treatments are differentiated into treatments that create an impervious surface, without pore filling effect (hydrophobic impregnation), treatments that make the surface less porous, with complete or incomplete pore filling effect (impregnation), and treatments that coat the concrete surface with an uninterrupted protective layer

DOI: 10.1201/9781003196143-8

(coatings). Typically relying on the application of silanes or siloxanes, hydrophobic impregnation is designed to slacken the process of deterioration by hindering the water from permeating the concrete, on the premise that, under water-free conditions, adverse chemical and physical–mechanical processes are extremely slow or insignificant [5–7]. Meanwhile, organic polymers are the basis on which impregnation and coatings develop a physical barrier to prevent permeation by harmful agents (e.g. chloride and other soluble salts, carbon dioxide, chemical agents, and water) causing concrete and/or reinforcement to break down. However, the employment of organic products is problematic in terms of durability, since such products are susceptible to various factors (e.g. UV, oxygen, temperature, humidity, pollutants), therefore suffering rapid loss of their original properties in outdoor settings. For instance, silanes, siloxanes, and polysiloxanes become less efficient after round 5 years, requiring regular application of surface treatment [8]. Organic products are also problematic in terms of compatibility, particularly in the case of application of continuous coatings or pore-blocking sealers, where water vapour transport is nearly non-existent and the protective surface layer may come off due to the water trapped behind it.

Persistent studies revealed that the nanoparticles of various construction materials can yield better effects as the filling agent compared to their micron-size counterparts. Guterrez [24] stated the feasibility of every material's conversion into its nanoparticles forms via grinding or treating chemically. The fabrication accurateness of these nanoparticles depends on the nature of the constituent parent materials and their purities. In the nanoscience and nanotechnology of material production at the nanoscale, the top-down [25] and bottom-up [26] approaches are followed. The selection of these two routes depends on the correctness, cost-effectiveness, and in-depth knowledge on the nanoscale characteristics of the materials under study [27]. In this view, the nanotechnology-enrouted materials received interests in the concrete research [28,29]. Undoubtedly, nanomaterials have shown notable promise to improve the mechanical strength and durability of the concretes/cements, thereby contributing towards the green construction materials' development [30,31].

Food, medicine, pharmaceutics, chemical and biological testing, manufacturing, construction materials, and coatings are just some of the sectors making extensive use of nanoparticles [15,32–36]. Evidence from a number of studies [37–39] supports the fact that nanomaterials can make polymer coatings for metal protection significantly stronger, tougher, more ductile, less permeable, and less susceptible to corrosion and wear. In a study on how resistant nanomaterials in latex paints were to weather, Feng et al. [40] observed that aging was prevented through UV ray absorption and scattering by nano-TiO_2 particles. In a different study, Xia et al. [41] examined how resistant graphene/epoxy composite coatings were to corrosion under γ-ray irradiation and found that the harmful effects of γ-ray irradiation were inhibited by graphene functioning as radical scavenger.

Another application of nanomaterials is in concrete coatings to make such coatings more durable. There is evidence that concrete protection is considerably enhanced by nano-clay in polymer paints, making concrete less susceptible to ingress of moisture, chloride ions, and surface water, salt attack, as well as colour alterations [19,42,43]. The implications of the addition of nano-SiO_2 or nano-TiO_2 for the property of epoxy resin (ER) coatings have been analysed by a few studies [31,44]. It was concluded that the addition of such materials made coated concrete more hydrophobic. The results of the carbonation experiments undertaken by Li et al. [45] on concrete with nano-SiO_2-modified polymer coatings in the aftermath of synthetic and natural weather aging indicated that concrete carbonation resistance remained effective for longer owing to the applied coatings.

In recent times, research has paid considerable attention to concrete surface treatments, with increasing importance being attributed to surface treatment technology in concrete structures, particularly in protecting such structures from breakdown due to exposure to harsh conditions as well as in improving their lifespan. Concrete treatments based on nanomaterials constitute the focus of the present chapter, with emphasis on surface coating, pore-blocking surface treatment, and multifunctional surface treatment. Furthermore, the manner in which nanomaterials interact with the cementitious substrate is examined. Nevertheless, the extensive investigations that have been conducted over the years have not fully elucidated the way in which many of the latest surface treatments work, particularly those that are based on nanomaterials. Hence, it is necessary to develop a more in-depth comprehension of the nano-scale mechanisms of chemical and physical reactions.

8.2 CONCRETE DURABILITY

Characteristically, the serviceability of construction materials has considerable economic significance, particularly with modern infrastructures and components. For urbanization, concrete materials that are greatly exploited must meet the standard codes of practice requisites related to strength and durability [46]. For instance, poor plan, low capacity or overload, faulty material design and structures, wrong construction practices or unsatisfactory maintenance, and lack of engineering knowledge can often diminish the service lifespan of concrete under operation [47]. In the construction industry, fast declination of concrete structures being a major setback necessitates additional improvement (Figure 8.1). Varieties of physical, chemical, thermal, and biological processes are responsible for the progressive deterioration concrete structures during their service [8,48]. Several studies [49–51] revealed that

FIGURE 8.1   Effect of aggressive environments on concrete durability performance [7].

the concrete performance is greatly affected by improper usage and physical and chemical conditions of environments. It is verified that both external and internal factors involving physical, chemical, or mechanical actions are often responsible for the deterioration of the concrete structures.

Mechanical damage of concrete structure occurs due to different reasons such as impact, abrasion, cracking, erosion, cavitation, or contraction. Chemical actions that cause the declination of concretes are carbonation, reaction associated with alkali and silica, alkali and carbonate as well as efflorescence. Moreover, outside attacks of chemicals happen primarily due to CO_2, Cl_2, SO_4 as well as several other liquids and gases generated by industries. Physical causes of deterioration include the effects of high temperature or differences in the thermal expansion of aggregate and of the hardened cement paste. Other reason of deterioration is the occurrence of alternating freezing and thawing of concrete and the associated action of the de-icing salts. Physical and chemical processes of deterioration often act in a synergistic way including the influence of sea water on concrete. Poor durability performance of OPC in aggressive acidic or sulphate (especially in marine) environment is caused due the existence of calcium complexes. These calcium complexes are very easily dissolved in acidic atmosphere, leading to enhanced porosity and thus fast deterioration [52].

In many places worldwide, OPC structures that existed for many decades are facing rapid deterioration [53]. Definitely, the permanence of OPC is linked with the nature of concretes' constituents, where CaO of 60%–65% and the hydration product of $Ca(OH)_2$ nearly 25% are responsible for fast structural decay. Several studies indicated that occurrence of fast reaction of $Ca(OH)_2$ in the acidic surroundings allows OPC to be deprived of water, leading to acid fusion and weakening of resistance against aggressive attacks. On the top, the intense reaction of evolved CO_2 with $Ca(OH)_2$ contributes to rapid corrosion of the concretes containing OPC [53]. The safety, service life, permanence, and life span of the mix design of concretes are considerably influenced by the crack development and subsequent erosion. These distinguished drawbacks of OPC-based concretes drove researchers to enhance properties of conventional OPC by adding pozzolanic materials, polymer, and nanomaterials so that it becomes more sustainable and endurable. The immediate consequence for affected concrete structures is the anticipated need of maintenance and execution of repairing [32]. Thus, there is a renewed interest for the development of sustainable concrete to solve all these existing shortcomings involving harsh environmental conditioning and durability.

The deterioration of concrete due to exposure to environment and mechanical loads requires continuous repair and rehabilitation. Among many degradation mechanisms, corrosion of steel rebar inside concrete is one of the most significant and detrimental. The chemical reaction in concrete produces calcium hydroxide, which provides an alkaline environment. Thus, a stable oxide film is formed on the steel surface, protecting the rebar from corrosion. However, the penetration of chloride ions, from sea water or de-icing salts, can break down the protective film and expose the steel bar to corrosion. To delay the chloride penetration and hence to prolong the service life of concrete structures, surface treatment such as coating is commonly employed [43].

8.3   COATING TECHNOLOGY IN CONCRETE

Defined by a desired reaction to certain external factors (e.g. temperature, light, humidity), the notion of "intelligent material" has emerged from the supplementary functionalities enabled by the latest advances in material technology. Successful development and assessment of sophisticated building materials (coatings) have been achieved, including coatings for concrete capable of repairing itself and having particular durability features [54,55]. Self-repair coatings are employed in reinforced concrete to enable steel bar self-repair and minimize corrosion damage. As a relatively new technology, self-repair coatings could provide a solution to the degradation affecting contemporary infrastructure [56]. Even slight damage to the coating can make standard anti-corrosion coatings less effective [57]. By contrast, self-repair coatings can repair the damage and carry on working at high efficiency, thus potentially helping to significantly extend the service life of steel rebar structures. Chen and colleagues [58] pioneered research on use of coatings on steel rebar. Self-repair coatings are deemed suitable for rebar structures in northeast regions with high levels of corrosion, where epoxy coatings are usually applied to slow down corrosion.

Numerous early studies have sought to gain insight into how to prevent the formation of minute cracks in concrete structures (e.g. roads, bridges), but this problem has not been fully and satisfactorily resolved [13,14,48]. Cracked concrete is susceptible to ingress by water, de-icing salt, and air. In winter, when salt is spread on roads, the concrete breaks down quicker because cracks are enlarged through the expansion of the frozen water within them. Metal protection based on self-repair anti-corrosion coatings has received ample attention, whereas the application of such coatings for concrete protection has not been investigated. A great deal of interest has been roused by self-repair coatings that enable cracked or damaged concrete to repair itself. Such coatings usually include micro-containers that burst readily when targeted. The healing agents in these containers enhance the service life of the coatings by sealing the cracks. The containers come in various forms, including polyurethane (PO) microcapsules and microfilament tubes and do not generally impact the mechanical properties of the coatings. Promising findings have been obtained by studies in this field, warranting additional exploration of use in real-world experiments [57,59,60].

To this end, a wide range of both organic and inorganic materials are employed. Rebar protection against corrosive agents (e.g. water, salts) is often achieved with epoxy coatings. Research has also been focused on creating polymer coatings capable of promoting crack repair by reacting to environmental factors like alterations in temperature and pH. Temperature-responsive coatings have been successfully developed and some have kept their mechanical properties even following repetitive repair cycles [61,62]. For instance, polyelectrolyte nanocontainer coatings can react to modifications in solution pH in a matter of seconds. This type of coating has great potential for future research because it can restore its mechanical properties completely [63].

Considerable interest has been raised by drying oils (e.g. tung oil, linseed oil) owing to their repair properties and encapsulation [64–66]. Exposure to air causes tung oil to undergo polymerization to an impervious coating characterized by toughness and glossiness, which is why paints, varnishes, and printing inks are heavily

FIGURE 8.2   Concrete coating technology [8].

reliant on drying oils. Samadzadeh and colleagues [67] were the first to accomplish tung oil encapsulation. The ASTM D4541 was applied to assess the pull-off strength of the urea-formaldehyde microcapsules, revealing that they adhered to the epoxy matrix more effectively than industry standards. Furthermore, evaluation of use life by introducing damaged specimens into solutions of sodium chloride yielded good outcomes, with use life being increased nine-fold by tung oil microcapsules compared to epoxy coatings following damage.

Concrete can be protected through different surface treatments (Figure 8.2), which can be generally distinguished as organic and inorganic, depending on the chemical structure of the agents they contain [68]. Organic surface treatments demonstrate satisfactory barrier properties, but their use life is suboptimal, whereas inorganic surface treatments exhibit greater stability and reduced susceptibility to aging, but their use is restricted. Another way of classifying surface treatments is based on functions: surface coatings, hydrophobic impregnation, pore-blocking treatments, and multifunctional surface treatments. Surface coating prevents damaging substances from entering concrete by producing an uninterrupted polymer film and generating a physical barrier [69,70]. Among the various surface coatings employed in foundations and quays are acrylic, butadiene copolymer, chlorinated rubber (CR), ER, oleoresinous, polyester resin, polyethylene copolymer, PO, vinyl, coal tar, and polymer-modified mortar [71]. Typically underpinned by water repellent products with a silane or siloxane basis, hydrophobic impregnation forms an impervious pore surface in the surface-near area without blocking the pores [43]. Meanwhile, pore-blocking surface treatments are designed to make the surface layer less porous by fully or partly filling the capillary pores [72]. Such treatments are usually performed with silicate-based pore blockers, but the latest pore-blocking products (e.g. nano-SiO_2 and $CaCO_3$ precipitation) are considered problematic. In recent times, pore-blocking treatment products have been used more and more often for the protection of buildings and highway bridges [73,74]. Last but not the least, multifunctional surface treatments employ products such as ethyl silicate and modified clay nanocomposites and are capable of filling the capillary pores as well as creation of a hydrophobic layer [42].

8.4   NANOTECHNOLOGY AND NANOPARTICLES

Compared to micron-sized materials, nanoparticles are accepted to have a better effect on filler. Figure 8.3 shows the available SiO_2 nanoparticles used for several applications in the concrete industry. As indicated by Guterrez [24], conversion of any material to nanoparticles can be achieved through crushing or chemical treatment, with the nanoparticle production precision depending on how pure the parent materials are and how they are chemically constituted. On a large scale, nanomaterials are typically generated via a top-down method [25] or a bottom-up method [26], with method selection being informed by suitability, cost, and the nanoscale properties of the material in question [27]. One example of a top-down method is milling, which is preferred because it is accessible, cost-effective, and readily permits alterations to be made with no reliance on chemical reagents or sophisticated electronic equipment [75]. The principle underlying the top-down method is conversion of a large structure (bulk) into a small structure (nano-dimension), with atomic-level control to prevent changes to the physical or chemical properties of the structure. This approach has

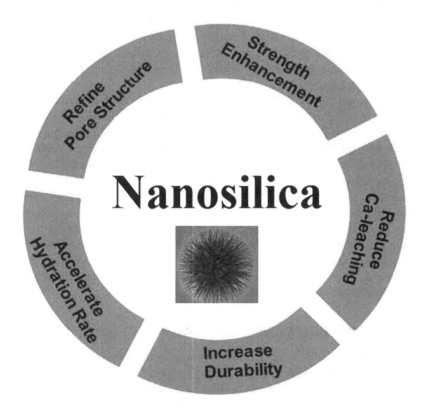

FIGURE 8.3 SiO_2 nanostructured material applications in the concrete industry.

been adopted at industrial level. A rich production of nanoparticles is made possible by the milling method. However, despite the compatibility of the top-down method with the purposes of nanofabrication, it does not allow reliable estimation of yield consistency and quality. One solution to this issue is to use a greater number of balls or types of balls, enhance the speed of milling, or change the character of the jar so as to improve nanoparticle quality [76].

A wide range of nanomaterials (e.g. nanoparticles, nanograins, nanoalloy, nano-composites, nano-quasicrystals) have been produced through high-energy ball mill-ing. This method was initially developed by John Benjamin for the fabrication of oxide particles within the nickel superalloy matrix. Milling permitted not only modi-fication of the properties of the alloy component to support high thermal structure but also improvement of mechanical strength. The milling-based transformation of a material into a particular morphological structure depends on fractures, plastic distortion/deformation, cold welds, and other factors. Besides crushing the material into small sections, milling involves the mixture of a large number of particles or materials for conversion into novel phases with various compositions [32]. In gen-eral, the milling end-products are flakes, which are refined according to the chosen ball and milling standard. The bottom-up method is the basis for the production of the majority of nanomaterials, including nano-silica, nanoalumina, and nanoclay, that are used in concrete. This method facilitates material engineering at atomic or molecular level through the self-assembly process known as molecular nanotechnol-ogy or molecular-level processing. Furthermore, the method is implemented in an indirect manner in nanomaterials and chemical production. Chemical synthesis is frequently performed to tailor the morphological features of the nanoparticles pro-duced through the bottom-up method.

The bottom-up method is better than the top-down method for producing more homogeneous and reproducible nanomorphology, whilst also facilitating the creation of new nanocrystals with ideal atomic or molecular arrangement. Furthermore, nano-materials fabricated via the bottom-up method display high electronic conductivity, optical absorption, and chemical reactivity. Surface atoms exhibiting consistency and minute size and associated with modified surface energies and morphologies can be achieved via the bottom-up method. Moreover, this method is useful for the pro-duction of nanomaterials capable of repairing and cleaning themselves and demon-strating superior catalytic properties, sensing capacity, and unique pigment features [77]. On the downside, the bottom-up method is expensive to run, requires expertise for chemical synthesis, and can only be applied in the laboratory. Nevertheless, the method yields nanoparticles that are ideal for electronic components, biotechnology, and other sophisticated applications [78].

8.5 NANOPARTICLES AND NANOMATERIALS

Consolidation nanoparticles have efficiency even in limited concentrations, which is why their properties have been the focus of a number of studies [79–81]. These materials have shown to have great potential in extreme conditions of frost, which they can effectively withstand. Empirical work on frost resistance has employed a wide range of nanoparticles in different concentrations, including nanoalumina, nanosilica, nanotitania, and other zero-dimensional particles [82,83], different forms of carbon nanotubes (CNTs) [84,85], nanoclay [86], and graphene [87].

It has been suggested that zero-dimensional nanoparticles are useful complementary additives that can serve as fillers, provide nucleation sites, improve hydrophobicity, and block empty spaces of middle size [88,89]. Besides this, some such nanoparticles (e.g. nanosilica [90,91], nanoalumina [90]) have a pozzolanic effect. Among the one-dimensional nanoparticles used in cementitious composites, CNTs have been advocated to have high efficiency, enhancing the hydration microstructure if they have a good distribution. CNTs have also been claimed to be pore structure contributors, diminishing the pore diameter from 200 to 50–100 nm, separating pores, and refining internal drainage, thus diminishing vulnerability to freeze-thaw cycles [92]. Meanwhile, with regard to other particles (e.g. graphene, nanoclay), no definitive conclusions can be formulated because they are yet to be fully developed. Nevertheless, there is enough evidence to suggest that compressive strength can be maintained with two-dimensional nano-additives through porosity decrease and pore rearrangement [87]. Furthermore, susceptibility to frost is diminished by both one- and two-dimensional particles through a mechanism of crack bridging that is discussed in the following chapter. Various proportions of 0.1%–7% of cement weight have been trialled and ideal results have been obtained with 2% nano-TiO_2, 5% nano-SiO_2, and 3% nano-Al_2O_3, which can respectively reduce strength loss by 12% [83], 16%, and 18% [82].

8.6   DEFINITION OF NANOTECHNOLOGY IN CONCRETE

The integration of nanotechnology into research on concrete is achieved via two major pathways known widely as nanoscience and nano-engineering or nano-modification of concrete. Employing sophisticated profiling methods and modeling at atomic or molecular level, nanoscience analyses the nano- and microscale structure of cement-based materials to gain insight into the impact of that structure on macroscale properties and performance [27,93,94]. Meanwhile, nano-engineering employs methods of nanoscale structure manipulation for the production of novel customized cementitious composites with multiple functionalities, high mechanical performance, and effective durability, as well as various new properties, including low electrical resistivity, self-sensing capacity, ability to clean and repair themselves, high ductility, and crack self-control. Nano-engineering of concrete involves regulation of how the material behaves and addition of new properties through the integration of nanoparticles and nanotubes or via molecule grafting on cement particles, phases, aggregates, and additives to achieve adjustable surface functionality aligned with particular interactions between interfaces.

8.7   NANOMATERIAL-BASED CONCRETES

In recent years, the growth in nanotechnology and the accessibility of nanomaterials suited for construction usage including nanosilica, nanoalumina, polycarboxylates, and nanokaolin have improved the concrete properties remarkably [27,95,96]. Intensive researches revealed that the mechanical properties such as compressive strength [33,34], splitting tensile and flexural strength of cement pastes [97,98], mortars [95,99], and concretes [100] can be improved via a tiny quantity of nanomaterials. Early strengths of pastes, mortars [95,99], and concrete [100] in the presence of nanomaterials were reported to be much higher than those formulated with conventional OPC. Development of such higher strengths was ascribed to the faster cement hydration process and pozzolanic reaction, reduction of pores density, and enhanced interfacial bonding amidst hardened cement paste and constituents (aggregates). Nanomaterials were also exploited to reduce the porosity and enhance the durability properties of concrete [101,102]. With developing the concrete technology, the nanomaterials were used in many applications such as ultra-high performance, self-cleaning, self-healing, fire protection, corrosion protection, and enhance the durability. The self-healing technology is one of the important applications of nanomaterials to produce sustainable and smart concrete.

8.8   PRODUCTION OF NANOCONCRETE

Inclusion materials with a particle size less than 500 nm in concrete production as admixture or part of cement replacement called nano-concrete. It was shown that the strength of normal concrete tends to enhance with the inclusion of nanoparticles. The bulk properties and packing model structure of concrete can remarkably be improved via the incorporation of nanoparticle. Nanoparticles act as excellent filling agents through the refinement of intersection zones in cementitious materials and production of high-density concrete. The manipulation or modification of these nanoparticles in the cement matrix can render a new-fangled nanostructures [103–105]. General deficiencies in the microstructures of concretes including voids, micro-porosity, and corrosion originated from the reaction of alkaline silica can be discarded. The advancement of nanomaterials occurred due to their characteristics as new binding agent with particle sizes much tinier than traditional OPC. This property enhances the hydration gel product by imparting a neat and solid structure. Besides, using a blend of filler and extra chemical reaction in the hydration scheme, high-performing novel nanoconcrete with enhanced durability can be achieved.

The application of nanotechnology in concrete is still in its infancy. Ever-growing demand for ultra-high performance concrete (UHPC) and recurring environmental pollution caused by OPC enforced the engineers to exploit nanotechnology in construction materials. Classical blends of UHPC with incorporated silica fumes can achieve enhanced strength and high durability. However, limited accessibility and high pricing of nanomaterials not only slowed down the growth of UHPC technology but also made it less demanding compared to conventional high strength concrete (HSC). To overcome these limitations, nanotechnology-enrouted production of UHPC emerged in its own right wherein an alternative to silica fume has been developed (Figure 8.4). Exploiting the nanoproduction idea, a typical nanomaterial mimicking the attributes of silica fume was designed. Nano-silica is certainly the newest material in nanotechnology-based processing that has been used as substitute to silica fume [106]. Using this celebrated nano-silica component, several types of nanoparticles have been synthesized which are effective for concrete production [107]. Nanoalumina [108], titanium oxide nanoparticles [109], CNT [110], and

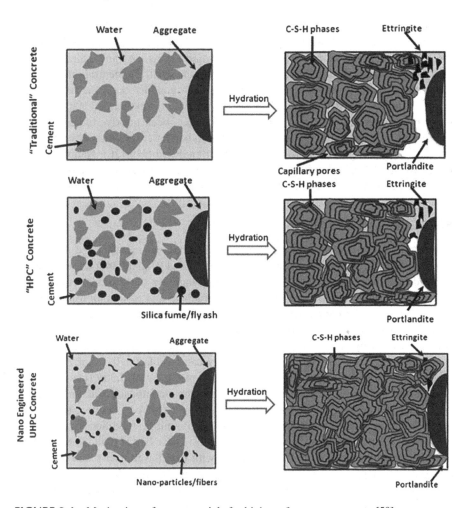

FIGURE 8.4   Mechanism of nanomaterials for high-performance concrete [59].

nanopolycarboxylates [111] are the emergent nanomaterials in the new nanoconcrete era. It is now customary to discuss about the production and possible applications of nanomaterials.

8.9   NANOMATERIAL-BASED CONCRETE SURFACE COATING

8.9.1   POLYMER NANOCOMPOSITE COATINGS

Compared to virgin polymers, polymer nanocomposite coatings display outstanding properties (e.g. superior strength, tensile modulus, abrasion and heat resistance, and thermal stability), which is why they have received attention from both research-ers and engineers. Gas permeability and flammability can be diminished by adding nanoparticles, while the barrier properties can be enhanced and the diffusion path improved, thus slackening polymer deterioration, through the addition of inorganic nanocomposites [112–115]. The diffusion path elongates with tortuosity, which rises with nanocomposite dosage and properties (e.g. aspect ratio, volume fraction) [116]. Unlike virgin polymer, nanocomposites can make polymer nanocomposite coatings 50–500 times less permeable to gas, even in small concentrations [19,117]. Nevertheless, only a handful of studies have been conducted on the use of such coatings in concrete structures. Likewise, the impact of polymer/clay nanocomposite coating properties on cementitious materials has been poorly studied [43,118]. Moreover, despite exhibiting promise regarding a barrier effect, the performance of polymer/SiO_2 and polymer/Al_2O_3 has not been analysed [119,120]. Nanocomposite materials can be cost-effectively developed with natural polymers (e.g. starch), which therefore deserve to be explored in terms of their use in concrete production [115].

8.9.2 Silane–Clay Nanocomposite Coatings

In the study by Woo et al. [43], the nano-clay material was incorporated to improve water repellence. Static water contact angles of 120°, 130°, and 142° were reported. It could reduce chloride penetration by 69% and increase static water contact angle to 142° [121]. It is obvious that the presence of silane and nanocomposite coatings significantly reduced the moisture permeability of concrete when applied on the surface. The hydrophobic nature of silane and coverage of micro voids present on the concrete surface by the coating along with the excellent barrier characteristics of clay with inherently large aspect ratio were mainly responsible for the sharply reduced moisture permeability. The barrier performance of the 5 wt% I.30P (primary octadecylamine modified, supplied by Nanocor Inc.) nanocomposite was only slightly better than the neat silane, whereas the barrier performance of 5 wt% Cloisite 20A (dimethyl dehydrogenated tallow quaternary ammonium modified, supplied by Rockwood Specialties) nanocomposite was almost twice better than the neat silane coating. This observation implies that I.30P with better wettability and lower viscosity does not necessarily improve the barrier performance of the coating more effectively than Cloisite 20A. The aspect ratio and dispersion of clay within the silane matrix, as well as the chemical interaction between the silane and different types of clay also contributed to different extents of moisture barrier performance.

Woo et al. [43] reported that the average chloride contents for the control specimen without coating and those with different types of coating were evaluated. The chloride contents were taken as an average over the depth up to 50 mm into the concrete of at least three specimens. The concrete without coating showed the highest chloride content of around 0.23 wt% of concrete mass. The chloride content dropped significantly with the application of coating: the neat silane and nanocomposite coatings brought an improvement with 92% and 69% reductions in chloride content, respectively. The effectiveness of silane as chloride barrier has also been reported previously based on different measurement techniques [122]. One of the main reasons for the better performance of the neat silane than the nanocomposite is that the silane coating is present deep into the concrete whereas the nanocomposite remains as a thin coating on the surface. It is likely that when the coating was applied onto the concrete surface, the neat silane coating could penetrate deeper into the concrete than the nanocomposite because of the much lower viscosity of silane and the absence of rigid fillers. The clay nanoparticles present in the thin nanocomposite surface coating did not contribute much to improving the barrier against chloride.

The barrier performance of neat silane and silane/clay nanocomposite coatings after accelerated weathering tests was evaluated. It was found that the micro voids present on the plain concrete surface were filled with coating materials. The elimination of micro voids due to coating means that the absorption and permeation of liquids or gases into the concrete can be effectively reduced. Both the neat silane and nanocomposite coatings significantly reduced the moisture permeability of concrete. The moisture barrier performance was better for the nanocomposites containing Cloisite 20A clay than those with I.30P clay or the neat silane coating. Approximately 5 wt% is found to be the optimal Cloisite 20A clay content that gave rise to the lowest permeability. The salt spray test indicates that the application of the neat silane and nanocomposite coatings reduced the average chloride content by 92% and 69%, respectively, compared to the uncoated concrete. The neat silane coating was able to permeate deeper into the concrete than the nanocomposite, indicating better chloride resistance performance of the neat silane.

8.9.3   SiO$_2$ AND/OR TiO$_2$ NANOPARTICLE COATING CONCRETE

SiO$_2$ and TiO$_2$ nanoparticles were used to produce high-performance concrete coating materials. Zhou et al. [123] found that with the incorporation of 1.5 wt% colloidal nanoparticles of TiO$_2$-SiO$_2$, the fabricated fluorocarbon/TiO$_2$-SiO$_2$ composite coatings exhibited a more stable hydrophilicity, superior self-cleaning, and anti-aging performance. Ammar et al. [124] studied the influence of nano-SiO$_2$ on hydrophobicity and corrosion resistance of acrylic–silicone polymeric matrix. They discovered that coatings incorporated with 3 wt% nano-SiO$_2$ exhibited the highest contact angle of 97.3° and significant improvements in corrosion resistance. Li et al. [125] investigated the influence of SiO$_2$ nanoparticles on hydrophobicity and carbonation resistance of concrete coated with polymer paints. They observed that coatings featured the highest surface contact angle and the best carbonation resistance when 1 wt% nano-SiO$_2$ was used. However, most current studies regarding application of nanomaterials in coatings focused on metal substrates [124,126], and a few studies centred on concrete substrates [125]. Characteristics of concrete and metal substrates differ.

Li et al. [44] used SiO$_2$ and TiO$_2$, it was found that the surface contact angles of concrete with PO, ER, or CR coating incorporated with different dosages of nano-SiO$_2$ or nano-TiO$_2$, as obtained from experimental results. Contact angles of concrete coated with PO, ER, or CR coatings totalled 72.4°, 68.9°, and 69.7° before introduction of nanoparticles, respectively, and their average value was nearly 70.3°. After addition of nanomaterials, contact angles of coated concretes all increased to different extents. Thus, addition of nanoparticles can effectively increase hydrophobicity of coating. However, given the increases in nanoparticle dosage, contact angles of coated concrete increased to a certain extent and then decreased. Such result coincides with the findings obtained by Ammar et al. [124] and Shafaamri et al. [127]. Possibly, with increasing dosage of nanoparticles, nanoparticle aggregation becomes increasingly serious [128], thereby causing additional difficulty in dispersion of nanoparticles in coatings. Ultimately, beneficial effects of nanoparticles will decrease or be lost. In corresponding dosages of nano-SiO_2 of 1.5, 2.5, and 0.5 wt% for PO, ER, and CR coating, contact angles of three types of coated concrete reached their maximum values of 87.2°, 85.2°, and 87.9°, respectively, which approached the critical value of 90° for hydrophobic materials. Compared with the coated concrete without nanoparticles, contact angles increased by 14.8°, 16.3°, and 18.2°, respectively. Average improvement value was 16.4°, and average relative improvement reached 23.4%. Similarly, in corresponding 2.5, 2, and 1 wt% dosages of nano-TiO_2 in PO, ER, and CR coatings, contact angles of the three types of coated concrete reached their maximum values at 87.3°, 80.2°, and 80.6°, respectively. Average improvement value was 12.4°, and average relative improvement amounted to 17.5%. Given the increasing nano-SiO_2 amount, contact angle values of three coatings stabilized, whereas those of nano-TiO_2 were highly discrete. In comparison with nano-TiO_2, nano-SiO_2 was slightly more effective in improving contact angles of the three types of coating.

The water absorption performance for each selected coated specimen with nanoparticles and blank specimens was assessed. Addition of nano-SiO_2 or nano-TiO_2 can reduce water absorption of coated concrete, confirming again that nanoparticles can improve hydrophobicity of coated concrete. Reduction ranges of water absorption for concrete coated with PO, CR, or ER coatings modified by nano-TiO_2 reached 17.9%, 12.3%, and 9.5%, respectively, with the average value totalling 13.2%. Similarly, such reduction ranges for concrete with PO, CR, and ER coatings modified by nano-SiO_2 amounted to 17.9%, 21.2%, and 14.4%, respectively, with the average value reaching 17.8%. Notably, nano-SiO_2 was more effective in decreasing water absorption of coated concrete than nano-TiO_2. Such result agrees with previous results on surface contact angles.

Nano-SiO_2 and nano-TiO_2 particles can be modified effectively from hydrophilicity into hydrophobicity by silane coupling agent KH-570, thereby guaranteeing dispersion stability of nanoparticles within organic film coatings. Nanoparticles can enhance contact angles on coated concrete by increasing surface roughness degree of coatings, whereas average maximum improvements by nano-SiO_2 and nano-TiO_2 totalled 23.4% and 17.5%, respectively. However, a dosage limit exists for nanoparticles in coatings, and improvements decrease once dosage exceeds the limit. Nanoparticles can reduce water absorption of coated concrete; average reduction by nano-SiO_2 and nano-TiO_2 totalled 17.8% and 13.2%, respectively. Water absorption of coated concrete is closely related with its contact angle, and wider contact angle lowers water absorption.

Li et al. [31] found that nano-TiO_2 can reduce microdefects in coating films and remarkably enhance the chloride resistance of coated concrete. Prior to coating aging, nano-TiO_2 can increase the chloride resistance of concrete with ER coatings by 76.8%, 61.8%, and 60%, respectively; nano-TiO_2 can effectively alleviate ultraviolet-induced damages on polymer coatings and thus evidently enhances the chloride resistance of coated concrete after ultraviolet radiation. After ultraviolet aging, nano-TiO_2 can reduce the Coulomb fluxes of concrete with CR, EP, and PO coatings by 32.2%, 40.2%, and 59.7%, respectively. Based on an S-shaped curve, models depicting the resistance of coated concrete to ultraviolet aging and chloride ions were established and used to predict the service lifetime. The results showed that nano-TiO_2 considerably extends the service lifetimes of polymer coatings.

Hou et al. [17] showed that the densification of the surface structure of cement-based material would be favourable for the improvement of the entire property of the bulk structure due to the inhibition of harmful substances in and outwards of the surface. In this study, effectiveness of colloidal nanoSiO$_2$ (CNS) with a mean particle size of 10 nm and its precursor, tetraethoxysilane (TEOS), for surface treatment of hardened cement mortar through brushing technique was investigated by studying their pozzolanic reactivity and pore-filling effect. Based on the study, the following conclusions can be drawn:

CNS is effective in decreasing water absorption ratio of cement mortar at 50°C and in sealed curing condition, but its effect at 20°C is negligible, especially in unsealed curing condition. Comparing with the CNS-treated sample, a grater reduction of water absorption can be obtained in TEOS-treated sample at 20°C or 50°C, in sealed or unsealed condition. Production of additional hydrates from the pozzolanic reaction between CNS/TEOS and Ca(OH)$_2$ and filler effects of CNS and TEOS contribute to the densification of the hardened cement-based materials. TEOS is more capable of filling pores finer than 50 nm, while CNS is effective in filling pores coarser than 50 nm, and the difference could be due to their difference in pore-penetration capability.

8.10   SUMMARY

In recent times, production of sustainable concrete via coating technology became useful in the construction industry worldwide. An all-inclusive overview of the appropriate literature on nanomaterial-based concrete coatings allowed us to draw the following conclusions:

   i. Nanomaterial-based concrete coatings are characterized by many significant traits such as ecofriendly and elevated durability performance in harsh environment. These properties make them effective sustainable materials in the construction industry.
   ii. Using nanomaterial-based concrete coating with improved performances and endurance useful for several applications is a new avenue in nanoscience and nanotechnology.
   iii. Incorporation of 4%–6% nanomaterials such as SiO$_2$, Al$_2$O$_3$ and TiO$_2$ into the concrete was found to increase the resistivity of the concrete and improved the corrosion behaviour of the concrete notably when exposed to seawater.
   iv. Use of nanomaterials in concrete coating is advantageous in terms of improved durability properties of cementitious materials, especially for the generation of self-healing and sustainable concretes.

REFERENCES

   1. Chen, R., et al., Effect of particle size of fly ash on the properties of lightweight insulation materials. *Construction and Building Materials*, 2016. **123**: pp. 120–126.
   2. Huseien, G.F., et al., Waste ceramic powder incorporated alkali activated mortars exposed to elevated Temperatures: Performance evaluation. *Construction and Building Materials*, 2018. **187**: pp. 307–317.

3. Mohammadhosseini, H., M.M. Tahir, and M. Sayyed, Strength and transport properties of concrete composites incorporating waste carpet fibres and palm oil fuel ash. *Journal of Building Engineering*, 2018. **20**: pp. 156–165.

4. Huseien, G.F., et al., The effect of sodium hydroxide molarity and other parameters on water absorption of geopolymer mortars. *Indian Journal of Science and Technology*, 2016. **9**(48): pp. 1–7.

5. Mhaya, A.M., et al., Long-term mechanical and durable properties of waste tires rubber crumbs replaced GBFS modified concretes. *Construction and Building Materials*, 2020. **256**: p. 119505.

6. Du, K., C. Xie, and X. Ouyang, A comparison of carbon dioxide (CO_2) emission trends among provinces in China. *Renewable and Sustainable Energy Reviews*, 2017. **73**: pp. 19–25.

7. Keyvanfar, A., et al., User satisfaction adaptive behaviors for assessing energy efficient building indoor cooling and lighting environment. *Renewable and Sustainable Energy Reviews*, 2014. **39**: pp. 277–295.

8. Huseien, G.F., et al., Geopolymer mortars as sustainable repair material: A comprehensive review. *Renewable and Sustainable Energy Reviews*, 2017. **80**: pp. 54–74.

9. Xie, T. and T. Ozbakkaloglu, Behavior of low-calcium fly and bottom ash-based geopolymer concrete cured at ambient temperature. *Ceramics International*, 2015. **41**(4): pp. 5945–5958.

10. Huseien, G.F., et al., Influence of different curing temperatures and alkali activators on properties of GBFS geopolymer mortars containing fly ash and palm-oil fuel ash. *Construction and Building Materials*, 2016. **125**: pp. 1229–1240.

11. Palomo, Á., et al., Railway sleepers made of alkali activated fly ash concrete. *Revista Ingeniería de Construcción*, 2011. **22**(2): pp. 75–80.

12. Huseiena, G.F., et al., Potential use coconut milk as alternative to alkali solution for geopolymer production. *Jurnal Teknologi*, 2016. **78**(11): pp. 133–9.

13. Ariffin, N.F., et al., Strength properties and molecular composition of epoxy-modified mortars. *Construction and Building Materials*, 2015. **94**: pp. 315–322.

14. Huseien, G.F., et al., Synthesis and characterization of self-healing mortar with modified strength. *Jurnal Teknologi*, 2015. **76**(1): pp. 195–200.

15. Huseien, G.F., K.W. Shah, and A.R.M. Sam, Sustainability of nanomaterials based self-healing concrete: An all-inclusive insight. *Journal of Building Engineering*, 2019. **23**: pp. 155–171.

16. Basheer, P., et al., Surface treatments for concrete: Assessment methods and reported performance. *Construction and Building Materials*, 1997. **11**(7–8): pp. 413–429.

17. Hou, P., et al., Effects and mechanisms of surface treatment of hardened cement-based materials with colloidal nanoSiO₂ and its precursor. *Construction and Building Materials*, 2014. **53**: pp. 66–73.

18. Asaad, M.A., et al., Improved corrosion resistance of mild steel against acid activation: Impact of novel Elaeis guineensis and silver nanoparticles. *Journal of Industrial and Engineering Chemistry*, 2018. **63**: pp. 139–148.

19. Scarfato, P., et al., Preparation and evaluation of polymer/clay nanocomposite surface treatments for concrete durability enhancement. *Cement and Concrete Composites*, 2012. **34**(3): pp. 297–305.

20. Asaad, M.A., et al., Rhizophora Apiculata as eco-friendly inhibitor against mild steel corrosion in 1 M HCL. *Surface Review and Letters*, 2017. **24**(Supp01): p. 1850013.

21. Bertolini, L., et al., *Corrosion of Steel in Concrete–Prevention, Diagnosis and Repair.* 2004, Wieley–VCH: Weinheim, pp. 1–12.

22. Brenna, A., et al., Long-term chloride-induced corrosion monitoring of reinforced concrete coated with commercial polymer-modified mortar and polymeric coatings. *Construction and Building Materials*, 2013. **48**: pp. 734–744.

23. Asaad, M.A., et al., Enhanced corrosion resistance of reinforced concrete: Role of emerging eco-friendly Elaeis guineensis/silver nanoparticles inhibitor. *Construction and Building Materials*, 2018. **188**: pp. 555–568.

24. Guterrez, K., How nanotechnology can change the concrete world. *American Ceramic Society Bulletin*, 2014. **84**(11), p. 17.

25. Abdoli, H., et al., Effect of high energy ball milling on compressibility of nanostructured composite powder. *Powder Metallurgy*, 2011. **54**(1): pp. 24–29.

26. Jankowska, E. and W. Zatorski. Emission of nanosize particles in the process of nanoclay blending. in *Third International Conference on Quantum, Nano and Micro Technologies*, ICQNM'09. 2009. IEEE: pp. 1–12.

27. Sanchez, F. and K. Sobolev, Nanotechnology in concrete–a review. *Construction and Building Materials*, 2010. **24**(11): pp. 2060–2071.

28. Aiswarya, S., A. Prince, and A. Narendran, Experimental investigation on concrete containing nano-metakaolin. *IRACST–Engineering Science and Technology: An International Journal*, 2013. **3**(1): pp. 2250–3498.

29. Kaur, M., J. Singh, and M. Kaur, Microstructure and strength development of fly ash-based geopolymer mortar: Role of nano-metakaolin. *Construction and Building Materials*, 2018. **190**: pp. 672–679.

30. Patel, K., The use of nanoclay as a constructional material. Department of Civil Engineering, LD College of Engineering Ahmedabad-Gujarat, India. *International Journal of Engineering Research and Applications*, 2012. **2**(4): pp. 1382–1386.

31. Li, G., et al., Improvements of Nano-TiO$_2$ on the long-term chloride resistance of concrete with polymer coatings. *Coatings*, 2019. **9**(5): p. 323.

32. Norhasri, M.M., M. Hamidah, and A.M. Fadzil, Applications of using nano material in concrete: A review. *Construction and Building Materials*, 2017. **133**: pp. 91–97.

33. Huseien, G.F., et al., Alkali-activated mortars blended with glass bottle waste nano powder: Environmental benefit and sustainability. *Journal of Cleaner Production*, 2019. **243**: p. 118636.

34. Samadi, M., et al., Influence of glass silica waste nano powder on the mechanical and microstructure properties of alkali-activated mortars. *Nanomaterials*, 2020. **10**(2): p. 324.

35. Cai, Y., et al., The effects of nanoSiO$_2$ on the properties of fresh and hardened cement-based materials through its dispersion with silica fume. *Construction and Building Materials*, 2017. **148**: pp. 770–780.

36. Zhang, B., et al., Nano-silica and silica fume modified cement mortar used as Surface Protection Material to enhance the impermeability. *Cement and Concrete Composites*, 2018. **92**: pp. 7–17.

37. Zhang, X., F. Wang, and Y. Du, Effect of nano-sized titanium powder addition on corrosion performance of epoxy coatings. *Surface and Coatings Technology*, 2007. **201**(16–17): pp. 7241–7245.

38. Wetzel, B., et al., Epoxy nanocomposites–fracture and toughening mechanisms. *Engineering Fracture Mechanics*, 2006. **73**(16): pp. 2375–2398.

39. Cheng, L., et al., Effect of graphene on corrosion resistance of waterborne inorganic zinc-rich coatings. *Journal of Alloys and Compounds*, 2019. **774**: pp. 255–264.

40. Li, F., et al., Research on dispersing properties of nano-materials in latex paints. *Journal of China University of Mining & Technology*, 2003. **6**: pp. 1–8.

41. Xia, W., et al., Graphene/epoxy composite coating damage under γ-ray irradiation and corrosion protection. *Journal of Inorganic Materials*, 2018. **33**: pp. 35–40.

42. Leung, C.K., et al., Use of polymer/organoclay nanocomposite surface treatment as water/ion barrier for concrete. *Journal of Materials in Civil Engineering*, 2008. **20**(7): pp. 484–492.

43. Woo, R.S., et al., Barrier performance of silane–clay nanocomposite coatings on concrete structure. *Composites Science and Technology*, 2008. **68**(14): pp. 2828–2836.
44. Li, G., et al., Influences of modified nanoparticles on hydrophobicity of concrete with organic film coating. *Construction and Building Materials*, 2018. **169**: pp. 1–7.
45. Li, G., et al., Long-term effectiveness of carbonation resistance of concrete treated with nano-SiO$_2$ modified polymer coatings. *Construction and Building Materials*, 2019. **201**: pp. 623–630.
46. Behfarnia, K., Studying the effect of freeze and thaw cycles on bond strength of concrete repair materials. *Asian Journal of Civil Engineering* 2010. **11**: pp. 165–172.
47. Gouny, F., et al., A geopolymer mortar for wood and earth structures. *Construction and Building Materials*, 2012. **36**: pp. 188–195.
48. Mirza, J., et al., Preferred test methods to select suitable surface repair materials in severe climates. *Construction and Building Materials*, 2014. **50**: pp. 692–698.
49. Mueller, H.S., et al., Design, material properties and structural performance of sustainable concrete. *Procedia Engineering*, 2017. **171**: pp. 22–32.
50. Husein, G.F., et al., Effects of POFA replaced with FA on durability properties of GBFS included alkali activated mortars. *Construction and Building Materials*, 2018. **175**: pp. 174–186.
51. Qureshi, T., A. Kanellopoulos, and A. Al-Tabbaa, Autogenous self-healing of cement with expansive minerals-II: Impact of age and the role of optimised expansive minerals in healing performance. *Construction and Building Materials*, 2019. **194**: pp. 266–275.
52. Chindaprasirt, P. and U. Rattanasak, Improvement of durability of cement pipe with high calcium fly ash geopolymer covering. *Construction and Building Materials*, 2016. **112**: pp. 956–961.
53. Hossain, M., et al., Durability of mortar and concrete containing alkali-activated binder with pozzolans: A review. *Construction and Building Materials*, 2015. **93**: pp. 95–109.
54. Pittaluga, M., The electrochromic wall. *Energy and Buildings*, 2013. **66**: pp. 49–56.
55. Gobakis, K., et al., Development and analysis of advanced inorganic coatings for buildings and urban structures. *Energy and Buildings*, 2015. **89**: pp. 196–205.
56. Lau, K. and A.A. Sagüés, Coating condition evaluation of epoxy coated rebar. *ECS Transactions*, 2007. **3**(13): pp. 81–92.
57. Shchukin, D.G., et al., Active anticorrosion coatings with halloysite nanocontainers. *The Journal of Physical Chemistry C*, 2008. **112**(4): pp. 958–964.
58. Chen, Y., et al., Self-healing coatings for steel-reinforced concrete. *ACS Sustainable Chemistry & Engineering*, 2017. **5**(5): pp. 3955–3962.
59. Cho, S.H., S.R. White, and P.V. Braun, Self-healing polymer coatings. *Advanced Materials*, 2009. **21**(6): pp. 645–649.
60. Nesterova, T., K. Dam-Johansen, and S. Kiil, Synthesis of durable microcapsules for self-healing anticorrosive coatings: A comparison of selected methods. *Progress in Organic Coatings*, 2011. **70**(4): pp. 342–352.
61. Yang, W.J., et al., Antifouling and antibacterial hydrogel coatings with self-healing properties based on a dynamic disulfide exchange reaction. *Polymer Chemistry*, 2015. **6**(39): pp. 7027–7035.
62. Luo, X. and P.T. Mather, Shape memory assisted self-healing coating. *ACS Macro Letters*, 2013. **2**(2): pp. 152–156.
63. Andreeva, D.V., et al., Self-healing anticorrosion coatings based on pH-sensitive polyelectrolyte/inhibitor sandwichlike nanostructures. *Advanced Materials*, 2008. **20**(14): pp. 2789–2794.
64. Suryanarayana, C., K.C. Rao, and D. Kumar, Preparation and characterization of microcapsules containing linseed oil and its use in self-healing coatings. *Progress in Organic Coatings*, 2008. **63**(1): pp. 72–78.

65. Jadhav, R.S., D.G. Hundiwale, and P.P. Mahulikar, Synthesis and characterization of phenol–formaldehyde microcapsules containing linseed oil and its use in epoxy for self-healing and anticorrosive coating. *Journal of Applied Polymer Science*, 2011. **119**(5): pp. 2911–2916.

66. Chen, Y., et al., Corrosion inhibition of mild steel in acidic medium by linseed oil-based imidazoline. *Journal of the American Oil Chemists' Society*, 2013. **90**(9): pp. 1387–1395.

67. Samadzadeh, M., et al., Tung oil: An autonomous repairing agent for self-healing epoxy coatings. *Progress in Organic Coatings*, 2011. **70**(4): pp. 383–387.

68. Franzoni, E., B. Pigino, and C. Pistolesi, Ethyl silicate for surface protection of concrete: performance in comparison with other inorganic surface treatments. *Cement and Concrete Composites*, 2013. **44**: pp. 69–76.

69. Pacheco-Torgal, F. and S. Jalali, Sulphuric acid resistance of plain, polymer modified, and fly ash cement concretes. *Construction and Building Materials*, 2009. **23**(12): pp. 3485–3491.

70. Diamanti, M.V., et al., Effect of polymer modified cementitious coatings on water and chloride permeability in concrete. *Construction and Building Materials*, 2013. **49**: pp. 720–728.

71. Bertolini, L., et al., Corrosion of steel in concrete. *Wiley Online Library*, 2013. **392**: pp. 1–3.

72. Dai, J.-G., et al., Water repellent surface impregnation for extension of service life of reinforced concrete structures in marine environments: the role of cracks. *Cement and Concrete Composites*, 2010. **32**(2): pp. 101–109.

73. Moon, H.Y., D.G. Shin, and D.S. Choi, Evaluation of the durability of mortar and concrete applied with inorganic coating material and surface treatment system. *Construction and Building Materials*, 2007. **21**(2): pp. 362–369.

74. Pan, X., et al., Effect of inorganic surface treatment on air permeability of cement-based materials. *Journal of Materials in Civil Engineering*, 2016. **28**(3): p. 04015145.

75. Shah, S.P., et al., Nanoscale modification of cementitious materials, in *Nanotechnology in Construction 3*, Bittnar, Z., Bartos, P.J.M., Němeček, J., Šmilauer, V., Zeman, J. (eds). 2009, Springer: Berlin, pp. 125–130.

76. Saleh, N.J., R.I. Ibrahim, and A.D. Salman, Characterization of nano-silica prepared from local silica sand and its application in cement mortar using optimization technique. *Advanced Powder Technology*, 2015. **26**(4): pp. 1123–1133.

77. Gesoglu, M., et al., Properties of low binder ultra-high performance cementitious composites: Comparison of nanosilica and microsilica. *Construction and Building Materials*, 2016. **102**: pp. 706–713.

78. Paul, K.T., et al., Preparation and characterization of nano structured materials from fly ash: Awaste from thermal power stations, by high energy ball milling. *Nanoscale Research Letters*, 2007. **2**(8): p. 397.

79. Ebrahimi, K., et al., A review of the impact of micro-and nanoparticles on freeze-thaw durability of hardened concrete: Mechanism perspective. *Construction and Building Materials*, 2018. **186**: pp. 1105–1113.

80. Korayem, A., et al., A review of dispersion of nanoparticles in cementitious matrices: Nanoparticle geometry perspective. *Construction and Building Materials*, 2017. **153**: pp. 346–357.

81. Zou, B., et al., Effect of ultrasonication energy on engineering properties of carbon nanotube reinforced cement pastes. *Carbon*, 2015. **85**: pp. 212–220.

82. Behfarnia, K. and N. Salemi, The effects of nano-silica and nano-alumina on frost resistance of normal concrete. *Construction and Building Materials*, 2013. **48**: pp. 580–584.

83. Salemi, N., K. Behfarnia, and S. Zaree, Effect of nanoparticles on frost durability of concrete. *Asian Journal of Civil Engineering*, 2014. **15**: pp. 411–420.

84. Kumar, S., et al., Effect of multiwalled carbon nanotube in cement composite on mechanical strength and freeze-thaw susceptibility. *Advances in Civil Engineering Materials*, 2015. **4**(1): pp. 257–274.

85. Li, W.-W., et al., Investigation on the mechanical properties of a cement-based material containing carbon nanotube under drying and freeze-thaw conditions. *Materials*, 2015. **8**(12): pp. 8780–8792.

86. Zhang, S., Y. Fan, and N. Li, Pore structure and freezing resistance of nanoclay modified cement based materials. *Materials Research Innovations*, 2014. **18**(2): pp. 358–362.

87. Tong, T., et al., Investigation of the effects of graphene and graphene oxide nanoplatelets on the micro-and macro-properties of cementitious materials. *Construction and Building Materials*, 2016. **106**: pp. 102–114.

88. Senff, L., et al., Effect of nano-silica on rheology and fresh properties of cement pastes and mortars. *Construction and Building Materials*, 2009. **23**(7): pp. 2487–2491.

89. Shekari, A. and M. Razzaghi, Influence of nano particles on durability and mechanical properties of high performance concrete. *Procedia Engineering*, 2011. **14**: pp. 3036–3041.

90. Salemi, N. and K. Behfarnia, Effect of nano-particles on durability of fiber-reinforced concrete pavement. *Construction and Building Materials*, 2013. **48**: pp. 934–941.

91. Rashad, A.M., A comprehensive overview about the effect of nano-SiO_2 on some properties of traditional cementitious materials and alkali-activated fly ash. *Construction and Building Materials*, 2014. **52**: pp. 437–464.

92. Parveen, S., S. Rana, and R. Fangueiro, A review on nanomaterial dispersion, microstructure, and mechanical properties of carbon nanotube and nanofiber reinforced cementitious composites. *Journal of Nanomaterials*, 2013. **2013**: p. 1–19.

93. Raki, L., J. Beaudoin, and R. Alizadeh, Nanotechnology applications for sustainable cement-based products, in *Nanotechnology in Construction 3*, Bittnar, Z., Bartos, P.J.M., Němeček, J., Šmilauer, V., Zeman, J. (eds). 2009, Springer: Berlin, pp. 119–124.

94. Scrivener, K.L. and R.J. Kirkpatrick, Innovation in use and research on cementitious material. *Cement and Concrete Research*, 2008. **38**(2): p. 128–136.

95. Li, H., et al., Microstructure of cement mortar with nano-particles. *Composites Part B: Engineering*, 2004. **35**(2): pp. 185–189.

96. Shah, K.W., et al., Aqueous route to facile, efficient and functional silica coating of metal nanoparticles at room temperature. *Nanoscale*, 2014. **6**(19): pp. 11273–11281.

97. Porro, A., et al., Effects of nanosilica additions on cement pastes. in *Applications of Nanotechnology in Concrete Design: Proceedings of the International Conference held at the University of Dundee, Scotland, UK on 7* July 2005. 2005. Thomas Telford Publishing: pp. 87–96.

98. Qing, Y., et al., Influence of nano-SiO_2 addition on properties of hardened cement paste as compared with silica fume. *Construction and Building Materials*, 2007. **21**(3): pp. 539–545.

99. Jo, B.-W., et al., Characteristics of cement mortar with nano-SiO_2 particles. *Construction and Building Materials*, 2007. **21**(6): pp. 1351–1355.

100. Schoepfer, J. and A. Maji, An investigation into the effect of silicon dioxide particle size on the strength of concrete. *Special Publication*, 2009. **267**: pp. 45–58.

101. Said, A.M. and M.S. Zeidan, Enhancing the reactivity of normal and fly ash concrete using colloidal nano-silica. *Special Publication*, 2009. **267**: pp. 75–86.

102. Zhang, M.-H. and J. Islam, Use of nano-silica to reduce setting time and increase early strength of concretes with high volumes of fly ash or slag. *Construction and Building Materials*, 2012. **29**: pp. 573–580.

103. Aydın, A.C., V.J. Nasl, and T. Kotan, The synergic influence of nano-silica and carbon nano tube on self-compacting concrete. *Journal of Building Engineering*, 2018. **20**: pp. 467–475.

104. Lim, N.H.A.S., et al., Microstructure and strength properties of mortar containing waste ceramic nanoparticles. *Arabian Journal for Science and Engineering*, 2018. **43**: pp. 1–9.

105. Fu, J., et al., Comparison of mechanical properties of CSH and portlandite between nano-indentation experiments and a modelling approach using various simulation techniques. *Composites Part B: Engineering*, 2018. **151**: pp. 127–138.

106. Yu, R., P. Spiesz, and H. Brouwers, Effect of nano-silica on the hydration and microstructure development of Ultra-High Performance Concrete (UHPC) with a low binder amount. *Construction and Building Materials*, 2014. **65**: pp. 140–150.

107. Adak, D., M. Sarkar, and S. Mandal, Effect of nano-silica on strength and durability of fly ash based geopolymer mortar. *Construction and Building Materials*, 2014. **70**: pp. 453–459.

108. Silva, J., et al., Influence of nano-SiO_2 and nano-Al_2O_3 additions on the shear strength and the bending moment capacity of RC beams. *Construction and Building Materials*, 2016. **123**: pp. 35–46.

109. Massa, M.A., et al., Synthesis of new antibacterial composite coating for titanium based on highly ordered nanoporous silica and silver nanoparticles. *Materials Science and Engineering: C*, 2014. **45**: pp. 146–153.

110. Morsy, M., S. Alsayed, and M. Aqel, Hybrid effect of carbon nanotube and nano-clay on physico-mechanical properties of cement mortar. *Construction and Building Materials*, 2011. **25**(1): pp. 145–149.

111. Navarro-Blasco, I., et al., Assessment of the interaction of polycarboxylate superplasticizers in hydrated lime pastes modified with nanosilica or metakaolin as pozzolanic reactives. *Construction and Building Materials*, 2014. **73**: pp. 1–12.

112. Pan, X., et al., A review on concrete surface treatment Part I: Types and mechanisms. *Construction and Building Materials*, 2017. **132**: pp. 578–590.

113. Woo, R.S., et al., Environmental degradation of epoxy-organoclay nanocomposites due to UV exposure: Part II residual mechanical properties. *Composites Science and Technology*, 2008. **68**(9): pp. 2149–2155.

114. Choudalakis, G. and A. Gotsis, Permeability of polymer/clay nanocomposites: A review. *European Polymer Journal*, 2009. **45**(4): pp. 967–984.

115. Fischer, H., Polymer nanocomposites: From fundamental research to specific applications. *Materials Science and Engineering: C*, 2003. **23**(6–8): pp. 763–772.

116. Kumar, A.P., et al., Nanoscale particles for polymer degradation and stabilization—trends and future perspectives. *Progress in Polymer Science*, 2009. **34**(6): pp. 479–515.

117. Ray, S.S. and M. Okamoto, Polymer/layered silicate nanocomposites: A review from preparation to processing. *Progress in Polymer Science*, 2003. **28**(11): pp. 1539–1641.

118. Carmona-Quiroga, P., et al., Interaction between two anti-graffiti treatments and cement mortar (paste). *Cement and Concrete Research*, 2010. **40**(5): pp. 723–730.

119. Woo, R.S., et al., Environmental degradation of epoxy–organoclay nanocomposites due to UV exposure. Part I: Photo-degradation. *Composites Science and Technology*, 2007. **67**(15–16): pp. 3448–3456.

120. Hackman, I. and L. Hollaway, Epoxy-layered silicate nanocomposites in civil engineering. *Composites Part A: Applied Science and Manufacturing*, 2006. **37**(8): pp. 1161–1170.

121. Muhammad, N.Z., et al., Waterproof performance of concrete: A critical review on implemented approaches. *Construction and Building Materials*, 2015. **101**: pp. 80–90.

122. Wittmann, F., T. Zhao, and H. Zhan. Establishment of an effective chloride barrier by water repellent impregnation. in *Proceedings of the International Workshop on Durability of Reinforced Concrete under Combined Mechanical and Climatic Loads (CMCL)*. 2005: pp. 105–117.

123. Zhou, J., et al., Preparation of transparent fluorocarbon/TiO_2-SiO_2 composite coating with improved self-cleaning performance and anti-aging property. *Applied Surface Science*, 2017. **396**: pp. 161–168.

124. Ammar, S., et al., A novel coating material that uses nano-sized SiO_2 particles to intensify hydrophobicity and corrosion protection properties. *Electrochimica Acta*, 2016. **220**: pp. 417–426.

125. Li, G., et al., Predicting carbonation depth for concrete with organic film coatings combined with ageing effects. *Construction and Building Materials*, 2017. **142**: pp. 59–65.

126. Bagherzadeh, M., A. Daneshvar, and H. Shariatpanahi, Novel water-based nanosiloxane epoxy coating for corrosion protection of carbon steel. *Surface and Coatings Technology*, 2012. **206**(8–9): pp. 2057–2063.

127. Ammar, S., et al., Amelioration of anticorrosion and hydrophobic properties of epoxy/PDMS composite coatings containing nano ZnO particles. *Progress in Organic Coatings*, 2016. **92**: pp. 54–65.

128. Zhang, Y. H., et al., Surface modification of nano-SiO_2 by silane coupling agent 3-(methacryloyloxy) propyltrimethoxysilane. *Journal of Materials Science and Engineering*, 2012. 5: pp. 1–8.

9 Nanomaterial-Based Concrete Antivirus Surface

9.1 INTRODUCTION

The primary means to combat the growth of the bacteria and viruses on various surfaces such as the glass, wood, or steel are to avoid their early adhesion onto these surfaces [1–3]. In this view, different types of coatings have been developed including the chlorhexidine-included hydroxyapatite, chlorhexidine-blended polylactide, polymers, and calcium phosphates plus chlorhexidine besides the nanomaterials such as copper (Cu) and silver (Ag) [1,4]. Diverse nanomaterials due to their unique and novel properties became useful for the biomedical, biosensing, optoelectronics, and catalysis applications [3]. These distinct characteristics of the nanomaterials emerge from their reduced size or dimension typically below 100 nm [5]. Depending on the structures and morphologies, these nanosystems are classified into zero dimensional (0D) such as the nanoparticles (NPs) and quantum dots (QDs); one-dimensional (1D) for example the nanofibres (NFs), nanotubes (NTs), nanorods (NRs) and nanowires (NWs); two dimensional (2D) including the quantum wells (QWs) and films such as the graphene and graphene oxide [6,7]; and three dimensional (3D) so called the bulk or macroscopic materials that are comprised of the equiaxed nanocrystallites with three random dimensions larger than nanometre [8]. In fact, some systems also exist in the boundaries of these classes of materials. In addition, the nanomaterials can be obtained naturally and prepared using various chemical, physical, biological, and mechanical techniques [6,9–11].

According to the International Organization for Standardization (ISO), each dimension of the NPs must be in the nanometre range. Conversely, for the NRs and nanoplates, the length along the largest and smallest direction is appreciably different [8]. The NPs can be prepared with different sizes, shapes, and dimensions such as the sphere, cylinder, cube, triangle, ring, or disk [12–14]. Generally, these NPs are classified depending on whether they are composed of carbon, organic, and inorganic compounds (Figure 9.1). The organic NPs are normally biodegradable without any toxicity including the dendrimer, ferritins, and hollow sphere (for instance, the micelle and liposome). The inorganic NPs are mostly based on different metals such as the silver (Ag), gold (Au), iron (Fe), aluminium (Al), cadmium (Cd), cobalt (Co), zinc (Zn), and Cu as well as the metal oxides including the TiO_2, FeO_2, Fe_3O_4, SiO_2, CeO_2, and ZnO_2. All these organic and inorganic nanomaterials reveal emerging properties those are completely absent in their bulk counterparts [15,16]. The carbon-based NPs can further be categorized into fullerene, graphene, and CNTs [17,18].

DOI: 10.1201/9781003196143-9

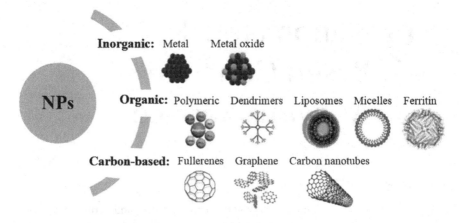

FIGURE 9.1   Classification of nanoparticles based on their nature.

The NPs being the linkages among the bulk structures as well as the atomic and molecular structures, their emergent novel characteristics mainly depend on the individual components. Consequently, the overall attributes of the NPs are appreciably different than the one existing at larger dimensions. Regardless of their nature, the most important properties that lead to the extensive uses of these NPs are related to their enlarged surface area, emergent optical traits due to the quantum size effects, improved absorbance, uniformities, and surface functionalizations. Especially, the effect of quantum confinement in the NPs leads to the emergence of the spontaneous semiconducting, conducting, or insulating properties for the adjacent particles [19], improved stabilities, and chemical and physical behaviours [17].

In recent times, the interests of using the NPs for varieties of biomedical applications have been ever-growing. Usually, the surfaces of the antimicrobial materials contain some chemical reagents responsible for inhibiting the microorganisms' growth [20]. Materials with such surfaces have widely been studied for various clinical, industrial, and domestic applications. The main uses of these materials are in the medical sectors as the antimicrobial coating and sterilizations of various equipment to prevent the spread of infection in hospitals that are responsible for numerous annual deaths in the United States [21]. Besides the medical equipment, the clothes are the ideal environments for the growth of many germs. Upon contacting the patients, these infected clothes enable an easy spread of these diseases [22]. To overcome such spread of the diseases, the antimicrobial surfaces have been functionalized through varieties of strategies. For instance, the applied coating to the surface may be the chemicals that have toxic effects to different microorganisms. Furthermore, via the attachments of various polymers or polypeptides on the frequently used surfaces, they can be functionalized to prevent the microbial infections [21].

An innovation in the antimicrobial surfaces began with the discovery of diverse nanomaterials such as Cu and its alloys (brass, bronze, cupronickel, copper–nickel–zinc, and others), Ag, Au, Ti, and Zn. These materials due to their unique antimicrobial traits can destroy varieties of microorganisms [23]. A recent review indicated the strong antimicrobial effectiveness of Cu against the *Escherichia coli* (O157:H7), methicillin-resistant *Staphylococcus aureus* (MRSA), *Staphylococcus*, *Clostridium difficile*, influenza A virus, adenovirus, and fungi [24]. Apart from the medical and healthcare sectors, the antimicrobial surfaces have also been used by other industries because of their self-cleaning traits. In fact, both physical and chemical characteristics of different surfaces can be customized to develop the antimicroorganism environments useful towards specific applications. Earlier, many photocatalytic materials have been utilized to kill several microorganisms. Lately, these photocatalytic materials became useful for the self-cleaning surfaces, air cleaning, water purification, and antitumor activity [25].

Considering the immense benefits of the nanomaterial-integrated antimicrobial surfaces, this article comprehensively overviewed the potential of several nanomaterials that can disinfect as well as deactivate viruses and pathogens transmitted through the air and contaminant surfaces. The antimicrobial efficacy of these nanomaterials is presented in terms of their types, surface morphologies, microstructures, fabrication procedures, and biocidal performances. In addition, an all-inclusive list of various nanomaterials is provided that are known to deactivate the deadly viruses on the common surfaces and protect humans against the infectious disease transmission. In short, the past developments, recent progress, and future prospects of these nanomaterials together with their environmental impact and biocidal effects and uses as the antiviral agents are underscored.

9.2   NANOMATERIAL-BASED ANTIMICROBIAL COATINGS

Any virus is comprised of a group of highly heterogeneous and simple organism with the size range of 100–300 nm (considerably smaller compared to the bacteria) [26]. The viruses do not have any independent metabolic activities and entirely depend on the living hosts for the reproduction. Unlike other living microorganisms, they enclose either DNA or RNA as the genetic components. The nuclear components in the viruses are surrounded by some layer of proteins for the protection from harmful agents present in the surroundings. In addition, such coating provides the docking sites required for attaching the viruses to their hosts. Some viruses possess an external cover composed of the lipid, polysaccharide, and protein molecules that enable the viruses to combine with the host cells, thus facilitating the entrance into the hosts. However, the viruses devoid of such external envelopes can also infect the host cell via other means. The viral diseases are commenced through the attachments of the special sites of the virus protein-coated surfaces on the specific receptor sites of the host cell surfaces [27,28].

The viral attacks can be inactivated by controlling the surfaces of the viruses particularly the docking sites, making the receptor site of the host cells totally unrecognizable. The coatings of diverse nanomaterials can effectively be used to inhibit the virus attack, thereby offering a viable solution to destroy the surface structures of the viruses. Currently, the antimicrobial, antiviral, and antifungal nano-coatings on the surfaces with various material compositions are available for the healthcare, household, and inside and outside uses that protect the decay and mildew together with water and air cleansing. In addition, the coatings of the nanomaterials are used to reduce the surface contaminants, improve the self-cleanliness, enhance the water-repellent tendency, improve the smell inhibition, and cost-saving maintenances. The antimicrobial, antiviral, and antifungal nano-coatings may be obtained via the spray, sinking, and attachments on various surfaces of the glasses, metals, alloys, marbles, stones, ceramic tiles, textiles, and plastics [29].

Generally, Covid-19 and other influenza viruses are transmitted among people via the airborne respiratory droplets escaped during coughing or sneezing. Such infections are spread when a healthy person comes in contact with these droplets that stick on the frequently used objects or surfaces [30]. Thus, an antiviral self-disinfecting surface can play a decisive role to destroy such viruses from further spreading. The glass surfaces coated with the hydrophobic (N, N-dodecyl, methyl-PEI) show strong antiviral traits against the lethal waterborne influenza A viruses including the wild-type human and avian strains in addition to their neuraminidase mutants resistant to the anti-influenza medicines [3]. In this regard, the nanotechnology and nanoscience with the immense biomedical potential can be effective for the self-disinfection of various surfaces. Being comparable to the nature, the combined nanoscience and biology can support to resist the pathogens, thereby offering an alternate strategy to combat against the contagious diseases [31].

Of late, the biological characteristics of various metallic NPs have been explored via the antimicrobial susceptibility tests. These NPs have been produced using different techniques. Some reports suggested that the metal NPs (silver, copper, copper oxides, and gold) have a broad range of the antimicrobial activities against diverse fungi and bacteria. The bactericidal action of these NPs against the *E. coli* [32] and *S. aureus* [33–35] has been reported. The uses of the proposed NPs were shown to inhibit these microorganisms growth. The metallic NPs generally disclose the antibacterial and antifungal activity despite the safety concern of the environment and human health related to the discharge and utilization of these NPs, which is an emerging area that needs further exploration. It was found that the surfaces coated with the nanomaterials of Cu, Ag, and Au can prevent the infection spread in the healthcare environments [36]. The microorganisms are known to survive on the inanimate surfaces for an extended time period [36]. Thus, the disinfection practices for the hands and surfaces became a primary measure against the microorganism-mediated infection spread. Approximately 80% of the infectious diseases are transmitted by touching the surfaces wherein the pathogens are found to survive on the inanimate surfaces for a prolonged time duration (days or months) in the healthcare facilities [37]. Thus, the microbial burden of the frequently touched surfaces plays a substantial role in the infection causalities [38]. In this perception, the nanomaterials can significantly address several clinical and public healthcare challenges emerged from Covid-19 pandemic. Based on these factors, we analyzed the usefulness of the nanotechnology and nanomaterials fighting against the coronavirus and ongoing mitigation strategies. The nanostructured material-based products are currently being developed and deployed for the inhibition, diagnosis, and treatment of Covid-19.

Following the nanotechnology route, Michielsen had modified the surfaces of the polymers/fibres using a thin dye coating [39]. This coating acted as the photocatalyst agent when exposed to the incident light, sparking a chemical reaction in the air that could kill most of the viruses and bacteria. A specific reaction occurred on the surface in the presence of light, making the air poisonous to the microbes without

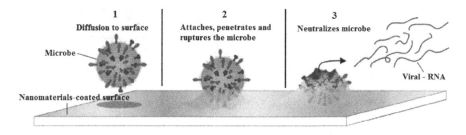

FIGURE 9.2 Working illustration of nanocoating technology [40].

causing any harm to human. In addition, the coating did not wear out and regenerated continually to kill the viruses (Figure 9.2). The Canadian researchers have developed a clear surface coating called the NanoClean that could kill the viruses (including the coronavirus) upon contact and lasted for several weeks [28]. This self-sterilizing nano-coating is among the latest Canadian technologies. Recent studies revealed that the SARS-CoV-2 that causes the Covid-19 can live on the surfaces for 72 hours and thus spread rapidly. The NanoClean was shown to kill up to 99.9% of the viruses and bacteria when applied on the surfaces. Moreover, its germ-fighting potency can be triggered by the light energy, thereby providing a longer protection against the surface-to-contact transmission than the conventional sterilizers. On top, it can be used on the high-touched surfaces such as the plastic chairs, doorknobs, and handrails to reduce the spread of Covid-19 [40].

9.3 TYPES OF NANOMATERIALS FOR ANTIMICROBIAL COATINGS

Over the years, various NPs have been used for different antimicrobial applications due to their distinctive characteristics [3]. Highly reactive nature of the nanomaterials offers their usage in the antimicrobial coatings [41]. Repeated studies indicated that various metallic and metal-oxide NPs such as the ZnO, CuO, Ag, CuI, Au-SiO as well as some quaternary ammonium cations are advantageous for the virucidal applications [42,43]. Amongst all the metal-based nanomaterials with the antimicrobial traits, the Ag and Cu NPs show the best antiviral activities [43,44]. The anti-HIV activity of the Ag NPs [45,46] was argued to be most likely due to their direct binding with the gp120 [47]. Meanwhile, the Ag NPs can hinder the growth of the viruses such as herpes simplex type 1 [48] and type 2 [49], vaccinia [50], respiratory syncytial [51], influenza A [52], tacaribe [53] and hepatitis B [47,54]. The upcoming sections describe the applications of various nanomaterials for the antiviral surface coatings.

9.3.1 COPPER (Cu)

The abundance of Cu and its various properties similar to the expensive noble metals (Ag and Au) make it a better choice for the antimicrobial applications. Recently, the Cu NPs generated renewed interests because of their catalytic, optical, and antimicrobial applications (Figure 9.3). The antimicrobial activities of the Cu NPs against several bacteria and fungi have been investigated [4,23]. Some studies indicated the

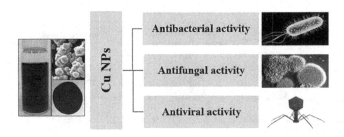

FIGURE 9.3   Schematic representation for the antimicrobial activity of Cu NPs [63].

antimicrobial activity of the Cu NPs against the *E. coli* and *Staphylococcus* species [55] with comparable antifungal attributes [56]. However, the rapid oxidation of the Cu NPs under the exposure of open air remains the major limitation. During the Cu NPs synthesis, the element Cu is oxidized to the CuO and Cu_2O, resulting in Cu^{2+} that makes the Cu NPs preparation difficult at an ambient environment. To overcome this problem, other strategies have been adopted to synthesize the Cu NPs wherein different polymers and surfactants are used as the stabilizing agents. In addition, the extracts from different plants have been utilized in the green synthesis techniques for stabilizing the as-prepared NPs [57]. The preparation of numerous types of NPs using the polymer dispersions has been reported [58]. Diverse techniques for the syntheses of the Cu NPs have been introduced including the thermal reduction [59], capping agent, sonochemical reduction, metal vapour [60], micro-emulsion [61], laser irradiation [62], and induced radiation.

The antimicrobial mechanisms of the Cu NPs have intensively been investigated [64]. The antimicrobial traits of the Cu NPs can be attributed to various mechanisms including:

 i. Elevated Cu levels inside a cell cause the oxidative stress and generation of the hydrogen peroxide, wherein Cu participates in the so-called Fenton-type chemical reaction responsible for the oxidative damage of the cells;
 ii. Excess Cu concentration causes the decay in the membrane integrity of the microbes, leading to the leaking of certain necessary cell nutrients such as the K and glutamine which in turn cause the desiccation and successive cell death;
 iii. For several performances the protein Cu is required; however, its excess amounts lead to the bindings of some proteins that are unnecessary for their function where such unwanted binding can lead to the improper working of these proteins and/or collapse into non-functional parts;
 iv. Continuous reduction of the bacterial attacks occurs, attaining 99.9% decrease within 2 hours of the application where more than 99.9% of the Gram-negative and Gram-positive bacteria are killed within 2 hours of the exposure. The presence of the Cu NPs delivers nonstop bactericidal activities on the residual living bacterial strains thereby destroying above 99.9% of them in 2 hours, killing over 99.9% of remaining bacterial strains in 2 hours and persist to destroy 99% of these strains still after the continual attacks;

v. It helps to inhibit the build-up and bacterial growth in 2 hours of the exposures between router cleaning and sanitization stage.

The antiviral efficacies of the Cu alloy surfaces have widely been investigated [65].
It was demonstrated that after the incubation for 1 hour on Cu, the effective influenza A viruses can be diminished by 75% and after 6 hours they can be reduced by
99.99% [66]. In addition, about 75% of the adenoviruses can be destroyed on the
Cu in 1 hour and 99.999% can be killed within 6 hours [67]. The Cu NPs in the size
range of 9.25 ± 1.79 nm were prepared using the wet-chemical process. The antifungal effectiveness of the Ag and Cu NPs was compared by measuring the inhibition zone diameter (IZD) using the disk-diffusion tests. The Ag NPs revealed higher
bactericidal effectiveness against the *E. coli* and *S. aureus* bacterial strains than the
Cu NPs. Conversely, the Cu NPs showed better antibacterial potency against the
Bacillus subtilis than the Ag NPs. The antifungal activity of the Ag and Cu NPs
against the Trametes versicolor fungus was examined [68]. Both NPs revealed considerable decrease in the mass loss, whereas the Cu NPs displayed better antifungal
effects than Ag NPs. In addition, the hardness values of two prepared panels were
dropped radically after the exposure to the fungi wherein the Cu NP-treated panels
revealed a significant improvement in the hardness. Several studies [69–72] have
been conducted to evaluate the antimicrobial potency of the commercial Cu NPs
against the *E. coli* and *B. subtilis* bacteria. The results showed a decrease in the
survival fraction of the bacteria with the increase in Cu NPs' contents. In short, the
replication of the bacteria was totally inhibited when the contents of Ag and Cu NPs
were above 70 and 60 lg/mL, respectively.

The antibacterial Cu NP coating can be used in the hospital, clinic, or home to
prevent the spread of the infections especially for the old people and infants at higher
risks. The antimicrobial effectiveness of the nano-Cu as additive in the architecture paints and impregnates was assessed [73]. The antimicrobial properties of the
nano-silica integrated with immobile Cu-NPs displayed the complete growth inhibition of the algae on the paints containing 0.5 ppm of the Cu-NPs was while control samples were intensively covered by the algae. In addition, the tested surfaces
covered with the paint containing Cu-NPs at the concentration of 0.1 ppm showed a
decrease in the microorganism growth by approximately 10%. The coatings containing the NPs of the safe metal ions and polymers displayed the excellent antiviral and
antimicrobial features that remained effective for weeks or even months [4]. The
antiviral coatings based on the Cu and other NP-embedded polymers were used to
paint/spray the surfaces. These NPs could control the release of the metal ions onto
the coated surface, displaying their strong antiviral effect via the eradication of the
virus particles that adhered to the surface. An extremely slow release of the ions in
the coating was greatly effective for the prolonged duration (over the weeks and even
months), inhibiting the virus particle's growth over ten-fold.

9.3.2 SILVER (Ag)

The antimicrobial traits of Ag are well known to the human since ages. The
Phoenicians stored water and other liquids in the silver-coated bottles to avoid the

contamination by the microbes. Silver dollars used to be put inside the milk bottles to keep the milk fresh. The water tanks of ships and airplanes were silvered to keep the water drinkable for months. Later, it was realized that amongst all antimicrobial metallic systems, the element Ag has highest bactericidal efficiency and lowest toxicity to the animal cells [74]. Earlier, it was believed that the atomic Ag binds to the enzymes' thiol group (-SH) and then deactivates them. The element Ag makes the S-Ag bond with the compound made of thiols responsible for the trans-membrane energy production and ion transportation in the cells [75]. In addition, the Ag atoms can participate in the catalytic oxidation reactions and generate the disulphide bonds (R-S-S-R) via the catalytic reactions among the O_2 in the cells and H in the thiols groups. In this process, H_2O is produced and two thiol groups make the covalent bonds via a disulphide linkage [76]. The Ag catalysis-mediated disulphide bond generation can alter the shapes of the enzymes in the cells and thereby affect their functions.

The Ag nanosystems are considered as the novel kinds of the bactericidal materials with excellent potency for the surface coating in the medical equipment, food packaging, and industrial piping. Therefore, the Ag NPs have intensively been investigated as the potential antimicrobial materials. The simple syntheses and strong antibacterial activities of the Ag NPs make them attractive for the future drug formulation. Martinez-Castanon et al. [77] prepared Ag NPs of different sizes (7, 29, and 89 nm) and determined the NPs' size-dependent antibacterial effectiveness. In the syntheses technique, Ag nitrate was used as the Ag ion source and gallic acid was utilized for the reduction and stabilization of the NPs. The NPs of various diameters were obtained by changing the solution's pH and UV irradiation energy. The TEM images were recorded to determine the NP morphologies and average diameters. The NP size-dependent bactericidal activities were assessed via the minimum inhibitory concentrations (MICs) assays against the *E. coli* (Gram negative) and *S. aureus* (Gram positive) bacterial strains. The smaller Ag NPs revealed stronger bacterial growth-inhibitory traits than the larger NPs wherein the *S. aureus* bacteria were more resistant to the Ag NPs than the *E. coli*. These findings agreed well with the observation made by Kawahara et al. [78]. The observed enhanced antimicrobial activity of the smaller NPs was attributed to their relatively easier penetration through the bacterial cell membrane and cell wall than the larger NPs [77]. Briefly, the tinier Ag NPs have higher number of atoms at the surface to get in touch with the solution compared to bigger ones. The results indicated that more Ag atoms from the smaller NPs were able to participate in the cell destruction processes compared to the larger Ag NPs. Assuming that only the atoms from the outer silver layer of the NPs were ionized, the larger NPs could produce less silver ions than the smaller one. In fact, the number of Ag ions was responsible for imparting the antibacterial properties to the NPs, thereby making the smaller Ag NPs more bactericidal than the larger Ag NPs. In addition to the sizes, the shapes of the NPs play the role to decide their bactericidal action.

Pal et al. [79] prepared the Ag NPs of different shapes (spherical, rod-like, and triangular) to determine their antimicrobial activity against the *E. coli* bacteria via the agar diffusion test, where the agar plates enclosed 1, 12.5, 50, and 100 μg of the above three kinds of Ag nanostructures as well as silver nitrate ($AgNO_3$). The numbers of the colonies produced under each condition were recorded and plotted against the Ag concentration for all the four types of treatments. The triangular Ag NPs showed the strongest antibacterial activity followed by the spherical Ag NPs, rod-shaped Ag NPs, and $AgNO_3$. The observed trend of the bactericidal potency of the Ag NPs and $AgNO_3$ was due to the presence of different facets of the NPs. The triangular Ag NPs had higher number of active faces (with dense electrons) compared to the spherical NPs, thereby showed better bactericidal efficacy than the spherical NPs. Conversely, the spherical Ag NPs facets were somewhat non-spherical but had higher number of the active faces compared to the rod-like NPs [79]. This clearly indicated the significant role of the Ag NPs with different shapes on the antibacterial activity. In short, the NPs with higher number of the active facets showed stronger bactericidal action.

The Ag powders with the ultra-fine and uniform particles size distributions are advantageous for the surface coating. Being a safe, non-toxic to animal cells, and effective bactericidal (highly toxic to the bacteria such as the *E. coli* and *S. aureas*), the metallic Ag became greatly demanding in all the technological sectors [80]. In recent years, various Ag-based compounds have broadly been used for the bacterial growth inhibition in the burn care applications [81]. In addition, the Ag-doped polymer fabrics, catheters, and polyurethane have been deployed for their notable antibacterial functionality [82]. Additionally, the colloidal silver [83], nano-Ag-coated fabric [1], nano-Ag metal oxide granules [84], and nano-Ag-coated ceramic materials [85] have been used for the antibacterial applications. Owing to their large surface area to the volume ratios, capacity of loading at small quantities, and cost-effectiveness, the nano-Ag powders and suspensions have extensively been used for widespread applications. Dedicated efforts have been made to produce monodispersed nano-sized Ag powders using different approaches. Pluym et al. [86] prepared nano-Ag powders via the spray pyrolysis technique at the production rate of 1–2 g/h. The produced silver NPs revealed agglomeration tendency, irregular shape, and hollow types due to the solvent evaporation.

It is established that the antibacterial potency of various nanomaterials is mainly because of their wide surface areas and diverse morphologies' (sizes and shapes) dependence of the physico-chemical characteristics [87–90]. Amongst different bactericidal nanosystems, the nano-Ag is considered to be the best because of its superior bactericidal potential compared to the bulk Ag [91,92]. Additionally, the nano-Ag has the wide range of bactericidal activates against various bacterial strains occurring in daily life, aggressive environment, and industrial processing such as the antibiotic-resistant bacteria [93]. Thus, it is essential to introduce the nanosilver as the antibacterial coating on various materials' surfaces for the widespread applications. Yet, the fabrication of the biocompatible coatings on the surfaces with high bactericidal effectiveness and environmental friendliness from the nano-Ag remains challenging [94,95]. The significant reduction in the antibacterial activities of the nanosilver can be ascribed to the aggregation of Ag NPs, uncontrollable release of the Ag ions, and improved adhesion of the bacterial strains [96,97].

Polymer-based compounds due to their excellent structural properties' customization and flexibility reveal strong inhibition potency to the agglomeration of the nano-Ag and formation of the homogeneous coating on the surfaces of various substrates [98]. In addition, such systems are able to regulate the discharge of the Ag ions for the constant bactericidal actions with low cytotoxicity [99]. Again, these materials can be tailored to inhibit the bacterial attachments, thereby enhancing the antibacterial effectiveness [100,101]. Therefore, it is beneficial to integrate the nano-Ag with the polymers to create the multi-functional coatings of the nanocomposites for the potential bactericidal applications. Accordingly, the polymer and nano-Ag nanocomposites having varied morphologies particularly the fibres were extensively investigated [102–104]. Recently, the antimicrobial properties and different functions of the Ag/polymer nanocomposites have been reviewed [105].

9.3.3 Silica (SiO2)

The nanotechnology may be used to fight against the microbial infections via the synthesis of various antimicrobial agents [106–108]. In this regard, silica is another inorganic and bio-friendly material used for the antiviral coating [109]. The extreme chemical stability and easy preparation of silica makes it widely popular with several notable advantages. The surface enrichment of the silica via the silanol groups enables easy reaction with the coupling agents, thereby providing strong attachment of the surface ligands on the metallic NPs (MNPs) [110]. In addition, the silica coating can enhance the stabilization of the MNPs in the liquid dispersions by preventing the dipolar attractions and increasing the number of surface charges, hence improving the electrostatic repulsions amid the particles in the non-aqueous dispersions. Numerous methods have been developed to create the silica coating. The Stöber method is the most popular one which is based on the hydrolysis reaction of the tetraethyl orthosilicate (TEOS) in the alcohol media under the catalysis by ammonia [111]. Following the Stöber method, Deng et al. [112] prepared silica-coated iron oxide core-shell NPs, where the condensation of the TEOS in the sol–gel form on the pre-formed magnetite NPs was used. The morphologies and coating thickness of these core-shell NPs was systematically tailored by adjusting the types and amounts of the alcohol, ammonia, and TEOS. Santra et al. [113] utilized the water-in-oil microemulsion method to coat the iron oxide NPs wherein different nonionic surfactants were used to obtain the NPs in the size range of 1–2 nm with uniform distribution (standard deviation below ±10%). Yi et al. [114] used a reverse microemulsion method to produce some homogenous silica-coated SiO_2/Fe_2O_3 NPs with various shell thicknesses. Finally, the mesoporous silica-coated SiO_2/Fe_2O_3 MNPs and hollow SiO_2 nanoballs were obtained [114]. The influence of coating thickness on the magnetic properties of the iron oxide NPs was determined. The silica-encapsulated Au-coated MNPs were attached with different functional components via the embedment or binding onto the silica shell to realize multifunctional nanosystems [115]. These NPs were claimed to be useful for various practical applications including the infectious disease treatment, cancer therapeutics, vaccine deliveries, and consumer food products [116,117].

9.3.4 Titanium (TiO2)

Usually, the decontamination of the surfaces is carried out repeatedly using the disinfectant solution to reduce the possibility of the recontamination. Recently, photocatalytic antimicrobial coating technology has been introduced as an alternative because of its long-lasting effect compared to the one-time use of the disinfectant solution [14]. The titanium dioxide or titania (TiO_2) has been used as the antimicrobial coating due to its strong photocatalytic properties [69,118]. The mechanisms of the TiO_2 photocatalysis involve the generation of the electron–hole (e^--h^+) pairs that diffuse and eventually get trapped on or near the TiO_2 surface in the presence of light [119]. The generated electrons and holes due to their strong reduction and oxidation potential can react easily with the atmospheric water and oxygen, creating hydroxyl radicals (OH^-) and superoxide anions (O_2^-). These radicals are extremely reactive to the organic compounds and bacterial cells, producing complete oxidation and depletion [119]. These properties of titania are exploited to disinfect the surfaces, air, food, and water as well as to produce anti-fogging and self-cleaning coatings on the glass and pharmaceutical additives [68]. The nanoscale TiO_2 has been claimed to be the excellent coating material for the antiviral, antibacterial, anti-static, anti-odour, reduction of the volatile organic compounds (VOCs), and UV radiation protection [3]. Some medical development divisions reported the strong antibacterial efficiency (in the range of 30%–95%) of the TiO_2 photocatalyst coatings against the *E. coli* and *S. aureus* [120]. Conversely, the TiO_2 photocatalyst is not that effective against some spore-forming bacteria (such as the *Bacillus atrophaeus*) and fungal species (e.g. *Aspergillus Niger* and yeasts).

Min et al. [121] conducted a prospective cohort study involving 621 patients in the medical intensive unit. The titania-based photocatalyst was coated on the high touch surface and walls. Some comparison of the multidrug-resistant organism incident rates, hospital-acquired blood stream infection, pneumonia, urinary tract infection, and *C. difficile* diseases was performed between the pre-intervention and post-intervention (5 months of data each set). The results revealed a significant decrease in the MRSA acquisition rate after the photocatalyst coating (hazard ratio of 0.37%, 95% confidence interval (CI) of 0.14 and 0.99 with p=0.04). However, the clinical identification of the vancomycin-resistant *Enterococcus* spp. and multidrug-resistant *Acinetobacter baumannii* did not display any significant reduction. The hazard of the hospital-acquired pneumonia during the intervention period compared to the baseline period was 0.46 (95% CI of 0.23 and 0.94 with p=0.03) [121].

Matthew et al. [122] examined the effectiveness of the TiO_2 coating to reduce the bioburden for the high-risk surfaces in between two acute care wards. The TiO_2 was sprayed onto six surfaces as coating agent and compared under the normal illumination against the same surfaces in an untreated ward with right and left bed rails, bed control, bedside locker, overbed table, and bed footboard. The general microbial burden and presence of an indicator pathogen (*S. aureus*) were evaluated biweekly for the period of 12 weeks. The titania-treated surfaces demonstrated a significant lowering of the microbial burden than the control sites wherein the difference between the treated and untreated surfaces was increased during the study. The hygiene failures (>2.5 colony-forming units [CFU]/cm^2) for the control surfaces were increased by 2.6% per day (odds ratio [OR] of 1.026, 95% CI of 1.009 and 1.043 with p=0.003). However, the hygiene failures for the treated surfaces were declined by 2.5% per day (with OR of 0.95, 95% CI of 0.925 and 0.977 with p<0.001) [123].

The coating materials with antimicrobial photosensitizers have advantages than the titania that is normally used as an antibacterial agent but ineffective against viruses. Following the approach of the Spontak team, both bacteria and viruses can be killed. The UV radiation is required to activate the titania that has detrimental effects on the healthy cells. Moreover, the use of the NPs has increasingly been scrutinized for the potential human health hazards during their exposures. The photoactivation of the titania by the UV radiation generates the reactive oxygen molecules on the TiO_2 surface that has been shown to effectively inactivate the influenza A virus [124], HIV-1 [125], and murine norovirus [126]. The halogen and interhalogen TiO_2 NPs, except for the chlorinated adduct, can completely inactivate the bacteriophages MS-2, ψ-X174, and PRD-1, displaying their oxidizing potential that can be generated without the UV photoactivation [127]. Similarly, the metal oxide NPs, CeO_2, Al_2O_3, and their halogen adducts exhibit excellent antiviral activities against the bacteriophages.

9.3.5 ZINC OXIDE (ZnO)

Generally, the semiconducting ZnO is accepted to be biocompatible without any toxicity to the human [128]. However, some of the recent reports revealed certain toxicity effects of the ZnO NPs due to its photocatalytic activity [129,130]. Amongst all the metal oxide NPs developed so far, the ZnO NPs display the highest toxicity against the microorganisms [131]. The SEM and TEM image analyses of the ZnO NPs showed that these NPs first damage the bacterial cell wall followed by the penetration and subsequent accumulation in the cell membrane. These NPs can interfere with the metabolic activities of the microbes, thereby leading to their deaths. Various properties of the ZnO NPs are decided by their morphologies, concentrations, and exposure times to the bacterial cells. The ZnO NPs absorb the UV radiation and promote the interactions with the molecules present in their immediate vicinity [132]. This photocatalytic process can continue long after their activation by the UV light absorption. The photocatalytic traits of the ZnO NPs can be attributed to the depletion of the surface electrons connected to the adsorbed negatively charged oxygen derivatives on the NPs surface. Upon activation with the UV light in the presence of oxygen, the ZnO NPs in the aqueous dispersions show a phototoxic action due to the formation of the reactive oxygen species (ROS). This in turn leads to the generation of the hydrogen peroxide (H_2O_2) that has a strong antimicrobial effect. Thereafter, the generated ROS can diffuse into the microbial cells and damage the interior as well as the cell walls, thus inhibiting their growth. The photocatalytic effect underpins the understanding of the antimicrobial action of the ZnO NPs in the drug formulations and other nanotechnology-based applications. The antimicrobial activity improvement of the ZnO NPs is due to the creation of more free radicals upon the activation by the UV light. Several researchers [130,133] have discussed the possible reaction mechanism of this effect. Nagarajan and Rajagopalan [134] established a correlation between the photon-initiated process of the photocatalytic NPs and their antimicrobial effect.

The photocatalytic titania has extensively been used due to its nontoxicity, high activity, strong self-cleaning traits, and low cost. The photocatalytic process of the

TiO_2 or ZnO involves the generation of the electron–hole pairs during the light exposure where the generated holes react with the nearby molecules to produce the oxidants. Then, the hydroxyl radicals (OH$^-$) are produced due to the chemical reaction of these particles with water and oxygen, creating the superoxide molecule. Similar to the titania, the ZnO and other semiconducting materials show unusual electronic and chemical properties useful for inhibiting the growth of the microorganisms [135–137]. Eventually, these radicals attack the bacteria or viruses via the inhibition of the DNA clonal processing, destroying the coenzymes and enzymes through the self-regeneration in the respiratory system. Consequently, the radical stops the reproduction of the bacteria and moulds, thereby inhibiting the bacteria growth or preventing the virus DNA multiplication [138,139].

The morphology of the ZnO nanostructures is strongly sensitive to the synthesis techniques. So far, diverse nanomorphologies of the semiconductor ZnO have been developed. These nanomorphologies include the NRs, nanoplates [140–142], nanospheres [143], nanoboxes [142], hexagonal, tripods [144], tetrapods [140], NWs, NTs, nanorings [145], nanocages, and nanoflowers [146]. Amongst all these nanostructures, the ZnO NPs are more active against the Gram-positive bacteria compared to other NPs made of the elements from the same group in the periodic table. Ready-to-eat food is more susceptible to the infection by *Salmonella*, *S. aureus*, and *E. coli* which pose a great challenge to the food safety and quality maintenance. Various antimicrobial compounds have been incorporated in the packed food to prevent them from the bacterial damage and decomposition. The antimicrobial packaging contains some nontoxic materials that inhibit or slow down the growth of the microbes present in the foods or packaging components [147].

The ZnO is one of the inorganic metal oxides that fulfil all the requirements of the antimicrobial food packaging. Thus, the ZnO NPs can safely be used as the medicine, preservative in packaging, and antimicrobial agent [148,149]. The easy diffusion of these NPs into the food products helps to kill the microbes and prevent the human being from falling ill. According to the 1935/2004/EC and 450/2009/EC regulations of the European Union, active packaging is defined as the active materials' contact with the food that can change the composition of the food or its surrounding atmosphere [150]. Therefore, it is commonly used as preservative and incorporated in the polymeric packaging material to prevent the microbes-related damage of the food materials [151]. The ZnO NPs have been used as the antibacterial substance against the *Salmonella typhi* and *S. aureus* in vitro.

The nanoscale ZnO has been demonstrated to exhibit diverse morphologies and significant antibacterial activity for a variety of bacterial species [152,153]. This wide band gap semiconductor material displays strong antimicrobial action when the particle dimension is decreased to the nanoscale. The nanoscale ZnO interacts strongly with the surfaces and/or with the cores of the bacteria by entering within the cell and reveals unique antibacterial actions [154]. The interaction amidst the unique nanostructures and bacterial strains is typically toxic, and thus suitable for the antimicrobial applications especially for the surface coatings. Interestingly, the ZnO NPs are nontoxic to the human cells [155] but toxic to the microorganisms. This distinct characteristic of the ZnO NPs enables their use as the bactericidal agents because of the noxiousness to the microorganisms and biocompatible nature to the human cells [134]. Different bactericidal mechanisms of various nanosystems are primarily ascribed to their large surface area to the volume ratio [156] and unique physico-chemical attributes. Briefly, in-depth studies of diverse nanomaterials are expected to strengthen our basic understanding on the antimicrobial mechanisms and fulfil the practical demands of the nanomaterials in the near future.

9.4 SUMMARY

Based on the critical survey of the relevant literatures, the following conclusions can be drawn:

  i. Various methods have been developed to fabricate different nanomaterial-based antiviral and antibacterial coatings.
 ii. The nanotechnology and nanomaterials became promising for customizing various properties of the materials desired for fighting the surface and air-borne pathogens, bacteria, and viruses.
iii. The manipulation of the materials' surface using various metal NP spray or coatings is essential for the long-term disinfection in an easy, economic, and yet safe manner.
 iv. The TiO$_2$ and ZnO NPs are the most common and efficient antimicrobial agents due to their bio-compatibility, non-toxicity, low cost, high chemical stability, and environmental responsively. Thus, the TiO$_2$ and ZnO nano-coatings have predominantly been used for the virucidal and bactericidal applications.
  v. The NPs of Cu Ag, TiO$_2$, SiO$_2$, and ZnO have shown high antimicrobial activity. Therefore, these nanomaterials can be used to fight against the biodeterioration.
 vi. It is asserted that diverse nanomaterials can be integrated into the existing building materials to improve their fundamental characteristics. In addition, the applications of different nanomaterial-based antiviral coatings may help to restrict or retard the spread of Covid-19.

REFERENCES

1. Mathew, T.V. and S. Kuriakose, Studies on the antimicrobial properties of colloidal silver nanoparticles stabilized by bovine serum albumin. *Colloids and Surfaces B: Biointerfaces*, 2013. **101**: pp. 14–18.
2. Peters Jr, T., *All about Albumin: Biochemistry, Genetics, and Medical Applications.* 1995, Academic Press: New York, pp. 1–18.
3. Shah, K.W. and W. Li, A review on catalytic nanomaterials for volatile organic compounds VOC removal and their applications for healthy buildings. *Nanomaterials*, 2019. **9**(6): p. 910.
4. Shah, K.W. and Y. Lu, Morphology, large scale synthesis and building applications of copper nanomaterials. *Construction and Building Materials*, 2018. **180**: pp. 544–578.
5. Malhotra, B.D. and M.A. Ali, Nanomaterials in biosensors: fundamentals and applications. *Nanomaterials for Biosensors*, 2018: p. 1.

6. Saleh, T.A. and V.K. Gupta, *Nanomaterial and Polymer Membranes: Synthesis, Characterization, and Applications.* 2016, Elsevier: Netherland, pp. 1–27.

7. Zhang, C., et al., Zero-dimensional, one-dimensional, two-dimensional and three-dimensional biomaterials for cell fate regulation. *Advanced Drug Delivery Reviews,* 2018. **132**: pp. 33–56.

8. Sudha, P.N., et al., Nanomaterials history, classification, unique properties, production and market, in *Emerging Applications of Nanoparticles and Architecture Nanostructures.* 2018, Elsevier: Netherland, pp. 341–384.

9. Huseien, G.F., et al., Alkali-activated mortars blended with glass bottle waste nano powder: environmental benefit and sustainability. *Journal of Cleaner Production,* 2019: p. 118636.

10. Hussein, A.A., et al., Physical, chemical and morphology characterisation of nano ceramic powder as bitumen modification. *International Journal of Pavement Engineering,* 2019: pp. 1–14.

11. Samadi, M., et al., Influence of glass silica waste nano powder on the mechanical and microstructure properties of alkali-activated mortars. *Nanomaterials,* 2020. **10**(2): p. 324.

12. Omrani, M.M., M. Ansari, and N. Kiaie, Therapeutic effect of stem cells and nano-biomaterials on Alzheimer's disease. *Biointerface Research in Applied Chemistry,* 2016. **6**(6): pp. 1814–1820.

13. Husain, Q., Nanosupport bound lipases their stability and applications. *Biointerface Research in Applied Chemistry,* 2017. **7**(6): pp. 2194–2216.

14. Nikolova, M.P. and M.S. Chavali, Metal oxide nanoparticles as biomedical materials. *Biomimetics,* 2020. **5**(2): p. 27.

15. Higa, A.M., et al., Ag-nanoparticle-based nano-immunosensor for anti-glutathione S-transferase detection. *Biointerface Research in Applied Chemistry,* 2016. **6**(1): 1053–1058.

16. Faisal, N. and K. Kumar, Polymer and metal nanocomposites in biomedical applications. *Biointerface Research in Applied Chemistry,* 2017. **7**(6): pp. 2286–2294.

17. Ealias, A.M. and M. Saravanakumar, A review on the classification, characterisation, synthesis of nanoparticles and their application. in *IOP Conference Series: Materials Science and Engineering,* 2017.

18. Teleanu, D.M., et al., Impact of nanoparticles on brain health: an up to date overview. *Journal of Clinical Medicine,* 2018. **7**(12): p. 490.

19. Strambeanu, N., et al., Nanoparticles: definition, classification and general physical properties, in *Nanoparticles' Promises and Risks.* 2015, Springer: Romania, pp. 3–8.

20. Dorland, N., Dorlands medical dictionary: antibacterial. Archived from the original on (17 November 2010), Retrieved (29 October 2010), 2010.

21. Onaizi, S.A. and S.S. Leong, Tethering antimicrobial peptides: current status and potential challenges. *Biotechnology Advances,* 2011. **29**(1): pp. 67–74.

22. Dastjerdi, R. and M. Montazer, A review on the application of inorganic nano-structured materials in the modification of textiles: focus on anti-microbial properties. *Colloids and Surfaces B: Biointerfaces,* 2010. **79**(1): pp. 5–18.

23. Usman, M.S., et al., Synthesis, characterization, and antimicrobial properties of copper nanoparticles. *International Journal of Nanomedicine,* 2013. **8**: p. 4467.

24. Marais, F., S. Mehtar, and L. Chalkley, Antimicrobial efficacy of copper touch surfaces in reducing environmental bioburden in a South African community healthcare facility. *Journal of Hospital Infection,* 2010. **74**(1): pp. 80–82.

25. Fujishima, M., et al., Isolation and structural analysis in vivo of newly synthesized fructooligosaccharides in onion bulbs tissues (Allium cepa L.) during storage. *International Journal of Carbohydrate Chemistry,* 2009. **2009**: pp. 1–10.

26. Vasilev, K., Nanoengineered antibacterial coatings and materials: a perspective. *Coatings*, 2019. **9**(10): p. 654.
27. Iwakoshi, A., T. Nanke, and T. Kobayashi, Coating materials containing gold nanoparticles. *Gold Bulletin*, 2005. **38**(3): pp. 107–112.
28. Phelan, A.L., R. Katz, and L.O. Gostin, The novel coronavirus originating in Wuhan, China: challenges for global health governance. *JAMA*, 2020. **323**(8): pp. 709–710.
29. Li, Y., et al., A smart multi-functional coating based on anti-pathogen micelles tethered with copper nanoparticles via a biosynthesis method using l-vitamin C. *RSC Advances*, 2018. **8**(33): pp. 18272–18283.
30. Fields, B., et al., *Fields' Virology*. 2001, Lippincott Williams & Wilkins: Philadelphia.
31. Janardhanan, R., et al., Synthesis and surface chemistry of nano silver particles. *Polyhedron*, 2009. **28**(12): pp. 2522–2530.
32. Chatterjee, A.K., et al., A simple robust method for synthesis of metallic copper nanoparticles of high antibacterial potency against *E. coli*. *Nanotechnology*, 2012. **23**(-8): p. 085103.
33. Cho, K.-H., et al., The study of antimicrobial activity and preservative effects of nanosilver ingredient. *Electrochimica Acta*, 2005. **51**(5): pp. 956–960.
34. Jung, W.K., et al., Antibacterial activity and mechanism of action of the silver ion in *Staphylococcus aureus* and *Escherichia coli*. *Applied and Environmental Microbiology*, 2008. **74**(7): pp. 2171–2178.
35. Abbasi, E., et al., Silver nanoparticles: synthesis methods, bio-applications and properties. *Critical Reviews in Microbiology*, 2016. **42**(2): pp. 173–180.
36. Arendsen, L.P., R. Thakar, and A.H. Sultan, The use of copper as an antimicrobial agent in health care, including obstetrics and gynecology. *Clinical Microbiology Reviews*, 2019. **32**(4): pp. e00125–18.
37. Kramer, A., I. Schwebke, and G. Kampf, How long do nosocomial pathogens persist on inanimate surfaces? A systematic review. *BMC Infectious Diseases*, 2006. **6**(1): p. 130.
38. Boyce, J.M., Environmental contamination makes an important contribution to hospital infection. *Journal of Hospital Infection*, 2007. **65**: pp. 50–54.
39. Michielsen, S., Surface modification of fibers via graft-site amplifying polymers. *International Nonwovens Journal*, 2003(3): p. 1558925003os–1200312.
40. Guo, L., et al., Polymer/nanosilver composite coatings for antibacterial applications. *Colloids and Surfaces A: Physicochemical and Engineering Aspects*, 2013. **439**: pp. 69–83.
41. Seh, Z.W., et al., Anisotropic growth of titania onto various gold nanostructures: synthesis, theoretical understanding, and optimization for catalysis. *Angewandte Chemie*, 2011. **123**(43): pp. 10322–10325.
42. Galdiero, S., et al., Silver nanoparticles as potential antiviral agents. *Molecules*, 2011. **16**(10): pp. 8894–918.
43. Hodek, J., et al., Protective hybrid coating containing silver, copper and zinc cations effective against human immunodeficiency virus and other enveloped viruses. *BMC Microbiology*, 2016. **16**(1): pp. 1–12.
44. Muñoz-Bonilla, A. and M. Fernández-García, Polymeric materials with antimicrobial activity. *Progress in Polymer Science*, 2012. **37**(2): pp. 281–339.
45. Lara, H.H., et al., PVP-coated silver nanoparticles block the transmission of cell-free and cell-associated HIV-1 in human cervical culture. *Journal of Nanobiotechnology*, 2010. **8**(1): p. 15.
46. Sun, R.W.-Y., et al., Silver nanoparticles fabricated in Hepes buffer exhibit cytoprotective activities toward HIV-1 infected cells. *Chemical Communications*, 2005(40): pp. 5059–5061.
47. Lara, H.H., et al., Mode of antiviral action of silver nanoparticles against HIV-1. *Journal of Nanobiotechnology*, 2010. **8**(1): pp. 1–10.

48. Baram-Pinto, D., et al., Inhibition of herpes simplex virus type 1 infection by silver nanoparticles capped with mercaptoethane sulfonate. *Bioconjugate Chemistry*, 2009. **20**(8): pp. 1497–1502.

49. Hu, R., et al., Inhibition effect of silver nanoparticles on herpes simplex virus 2. *Genetics and Molecular Research*, 2014. **13**(3): pp. 7022–7028.

50. Trefry, J.C. and D.P. Wooley, Silver nanoparticles inhibit vaccinia virus infection by preventing viral entry through a macropinocytosis-dependent mechanism. *Journal of Biomedical Nanotechnology*, 2013. **9**(9): pp. 1624–1635.

51. Sun, L., et al., Silver nanoparticles inhibit replication of respiratory syncytial virus. *Journal of Biomedical Nanotechnology*, 2008. **4**(2): pp. 149–158.

52. Xiang, D.-x., et al., Inhibitory effects of silver nanoparticles on H1N1 influenza A virus in vitro. *Journal of Virological Methods*, 2011. **178**(1–2): pp. 137–142.

53. Speshock, J.L., et al., Interaction of silver nanoparticles with Tacaribe virus. *Journal of Nanobiotechnology*, 2010. **8**(1): pp. 1–9.

54. Lu, L., et al., Silver nanoparticles inhibit hepatitis B virus replication. *Antiviral Therapy*, 2008. **13**(2): p. 253.

55. Ren, G., et al., Characterisation of copper oxide nanoparticles for antimicrobial applications. *International Journal of Antimicrobial Agents*, 2009. **33**(6): pp. 587–590.

56. Wei, Y., et al., Synthesis of stable, low-dispersity copper nanoparticles and nanorods and their antifungal and catalytic properties. *The Journal of Physical Chemistry C*, 2010. **114**(37): pp. 15612–15616.

57. Shameli, K., et al., Green biosynthesis of silver nanoparticles using Callicarpa maingayi stem bark extraction. *Molecules*, 2012. **17**(7): pp. 8506–8517.

58. Usman, M.S., et al., Copper nanoparticles mediated by chitosan: synthesis and characterization via chemical methods. *Molecules*, 2012. **17**(12): pp. 14928–14936.

59. Salavati-Niasari, M. and F. Davar, Synthesis of copper and copper (I) oxide nanoparticles by thermal decomposition of a new precursor. *Materials Letters*, 2009. **63**(3–4): pp. 441–443.

60. Ponce, A.A. and K.J. Klabunde, Chemical and catalytic activity of copper nanoparticles prepared via metal vapor synthesis. *Journal of Molecular Catalysis A: Chemical*, 2005. **225**(1): pp. 1–6.

61. Ohde, H., F. Hunt, and C.M. Wai, Synthesis of silver and copper nanoparticles in a water-in-supercritical-carbon dioxide microemulsion. *Chemistry of Materials*, 2001. **13**(11): pp. 4130–4135.

62. Giusti, A., et al., Multiphoton fragmentation of PAMAM G5-capped gold nanoparticles induced by picosecond laser irradiation at 532 nm. *The Journal of Physical Chemistry C*, 2007. **111**(41): pp. 14984–14991.

63. Ingle, A.P., N. Duran, and M. Rai, Bioactivity, mechanism of action, and cytotoxicity of copper-based nanoparticles: a review. *Applied Microbiology and Biotechnology*, 2014. **98**(3): pp. 1001–1009.

64. Ruparelia, J.P., et al., Strain specificity in antimicrobial activity of silver and copper nanoparticles. *Acta Biomaterialia*, 2008. **4**(3): pp. 707–716.

65. Michels, H.T., et al., From laboratory research to a clinical trial: copper alloy surfaces kill bacteria and reduce hospital-acquired infections. *HERD: Health Environments Research & Design Journal*, 2015. **9**(1): pp. 64–79.

66. Noyce, J., H. Michels, and C. Keevil, Inactivation of influenza A virus on copper versus stainless steel surfaces. *Applied and Environmental Microbiology*, 2007. **73**(8): pp. 2748–2750.

67. Karpanen, T., et al., The antimicrobial efficacy of copper alloy furnishing in the clinical environment: a crossover study. *Infection Control & Hospital Epidemiology*, 2012. **33**(1): pp. 3–9.

68. Taghiyari, H.R., et al., Effects of silver and copper nanoparticles in particleboard to control Trametes versicolor fungus. *International Biodeterioration & Biodegradation*, 2014. **94**: pp. 69–72.

69. Makvandi, P., et al., Metal-based nanomaterials in biomedical applications: antimicrobial activity and cytotoxicity aspects. *Advanced Functional Materials*, 2020: p. 1910021.

70. Kruk, T., K. Szczepanowicz, J. Stefan, R.P. Socha, and P. Warszyn, Synthesis and antimicrobial activity of monodisperse copper nanoparticles. *Colloids and Surfaces: B: Biointerfaces*, 2015. **128**(1): p. 17.

71. Rodríguez-Sánchez, I.J., et al., Electrospinning of ultra-thin membranes with incorporation of antimicrobial agents for applications in active packaging: a review. *International Journal of Polymeric Materials and Polymeric Biomaterials*, 2020: pp. 1–24.

72. Ferdous, Z. and A. Nemmar, Health impact of silver nanoparticles: a review of the biodistribution and toxicity following various routes of exposure. *International Journal of Molecular Sciences*, 2020. **21**(7): p. 2375.

73. Zielecka, M., et al., Antimicrobial additives for architectural paints and impregnates. *Progress in Organic Coatings*, 2011. **72**(1–2): pp. 193–201.

74. Guggenbichler, J.-P., et al., A new technology of microdispersed silver in polyurethane induces antimicrobial activity in central venous catheters. *Infection*, 1999. **27**(1): pp. S16–S23.

75. Klueh, U., et al., Efficacy of silver-coated fabric to prevent bacterial colonization and subsequent device-based biofilm formation. *Journal of Biomedical Materials Research: An Official Journal of The Society for Biomaterials, The Japanese Society for Biomaterials, and The Australian Society for Biomaterials and the Korean Society for Biomaterials*, 2000. **53**(6): pp. 621–631.

76. Davies, R.L. and S.F. Etris, The development and functions of silver in water purification and disease control. *Catalysis Today*, 1997. **36**(1): pp. 107–114.

77. Martínez-Castañon, G.-A., et al., Synthesis and antibacterial activity of silver nanoparticles with different sizes. *Journal of Nanoparticle Research*, 2008. **10**(8): pp. 1343–1348.

78. Kawahara, K., et al., Antibacterial effect of silver-zeolite on oral bacteria under anaerobic conditions. *Dental Materials*, 2000. **16**(6): pp. 452–455.

79. Pal, S., Y.K. Tak, and J.M. Song, Does the antibacterial activity of silver nanoparticles depend on the shape of the nanoparticle? A study of the gram-negative bacterium *Escherichia coli*. *Applied and Environmental Microbiology*, 2007. **73**(6): pp. 1712–1720.

80. Zhao, G. and S.E. Stevens, Multiple parameters for the comprehensive evaluation of the susceptibility of *Escherichia coli* to the silver ion. *Biometals*, 1998. **11**(1): pp. 27–32.

81. Klasen, H., Historical review of the use of silver in the treatment of burns. I. Early uses. *Burns*, 2000. **26**(2): pp. 117–130.

82. Cho, J.W. and J.H. So, Polyurethane–silver fibers prepared by infiltration and reduction of silver nitrate. *Materials Letters*, 2006. **60**(21–22): pp. 2653–2656.

83. Prucek, R., L. Kvítek, and J. Hrbáč, Silver colloids-methods of preparation and utilization. Acta Universitatis Palackianae Olomucensis Facultas Rerum Naturalium. *Nature Mathematica*, 2004. **43**: pp. 56–68.

84. Yan, J. and J. Cheng, Nanosilver-containing antibacterial and antifungal granules and methods for preparing and using the same, 2002, Google Patents.

85. Kawasumi, S., M. Yamada, and M. Honma, Metallic bactericidal agent, 1998, Google Patents.

86. Pluym, T., et al., Solid silver particle production by spray pyrolysis. *Journal of Aerosol Science*, 1993. **24**(3): pp. 383–392.

87. Mauter, M.S., et al., Antifouling ultrafiltration membranes via post-fabrication grafting of biocidal nanomaterials. *ACS Applied Materials & Interfaces*, 2011. **3**(8): pp. 2861–2868.

88. Duncan, T.V., Applications of nanotechnology in food packaging and food safety: barrier materials, antimicrobials and sensors. *Journal of Colloid and Interface Science*, 2011. **363**(1): pp. 1–24.

89. Rai, M., et al., Silver nanoparticles: the powerful nanoweapon against multidrug-resistant bacteria. *Journal of Applied Microbiology*, 2012. **112**(5): pp. 841–852.

90. Sharma, V.K., R.A. Yngard, and Y. Lin, Silver nanoparticles: green synthesis and their antimicrobial activities. *Advances in Colloid and Interface Science*, 2009. **145**(1–2): pp. 83–96.

91. Huda, S., et al., Antibacterial nanoparticle monolayers prepared on chemically inert surfaces by cooperative electrostatic adsorption (CELA). *ACS Applied Materials & Interfaces*, 2010. **2**(4): pp. 1206–1210.

92. Liu, J., et al., Controlled release of biologically active silver from nanosilver surfaces. *ACS Nano*, 2010. **4**(11): pp. 6903–6913.

93. Liong, M., et al., Antimicrobial activity of silver nanocrystals encapsulated in mesoporous silica nanoparticles. *Advanced Materials*, 2009. **21**(17): pp. 1684–1689.

94. Agarwal, A., et al., Surfaces modified with nanometer-thick silver-impregnated polymeric films that kill bacteria but support growth of mammalian cells. *Biomaterials*, 2010. **31**(4): pp. 680–690.

95. Lischer, S., et al., Antibacterial burst-release from minimal Ag-containing plasma polymer coatings. *Journal of the Royal Society Interface*, 2011. **8**(60): pp. 1019–1030.

96. Oei, J.D., et al., Antimicrobial acrylic materials with in situ generated silver nanoparticles. *Journal of Biomedical Materials Research Part B: Applied Biomaterials*, 2012. **100**(2): pp. 409–415.

97. Lee, D., M.F. Rubner, and R.E. Cohen, Formation of nanoparticle-loaded microcapsules based on hydrogen-bonded multilayers. *Chemistry of Materials*, 2005. **17**(5): pp. 1099–1105.

98. Wang, Y., et al., Templated synthesis of single-component polymer capsules and their application in drug delivery. *Nano Letters*, 2008. **8**(6): pp. 1741–1745.

99. Silva, A.R. and G. Unali, Controlled silver delivery by silver–cellulose nanocomposites prepared by a one-pot green synthesis assisted by microwaves. *Nanotechnology*, 2011. **22**(31): p. 315605.

100. Li, P., et al., Antimicrobial macromolecules: synthesis methods and future applications. *RSC Advances*, 2012. **2**(10): pp. 4031–4044.

101. Timofeeva, L. and N. Kleshcheva, Antimicrobial polymers: mechanism of action, factors of activity, and applications. *Applied Microbiology and Biotechnology*, 2011. **89**(3): pp. 475–492.

102. Kong, H. and J. Jang, Antibacterial properties of novel poly (methyl methacrylate) nanofiber containing silver nanoparticles. *Langmuir*, 2008. **24**(5): pp. 2051–2056.

103. Rujitanaroj, P.-o., N. Pimpha, and P. Supaphol, Wound-dressing materials with antibacterial activity from electrospun gelatin fiber mats containing silver nanoparticles. *Polymer*, 2008. **49**(21): pp. 4723–4732.

104. Yuan, J., et al., Electrospinning of antibacterial poly (vinylidene fluoride) nanofibers containing silver nanoparticles. *Journal of Applied Polymer Science*, 2010. **116**(2): pp. 668–672.

105. Dallas, P., V.K. Sharma, and R. Zboril, Silver polymeric nanocomposites as advanced antimicrobial agents: classification, synthetic paths, applications, and perspectives. *Advances in Colloid and Interface Science*, 2011. **166**(1–2): pp. 119–135.

106. Zare, E.N., et al., Metal-Based nanostructures/PLGA nanocomposites: antimicrobial activity, cytotoxicity, and their biomedical applications. *ACS Applied Materials & Interfaces*, 2019. **12**(3): pp. 3279–3300.

107. Zare, E.N., et al., Advances in biogenically synthesized shaped metal-and carbon-based nanoarchitectures and their medicinal applications. *Advances in Colloid and Interface Science*, 2020: p. 102236.

108. Jamaledin, R., et al., Advances in antimicrobial microneedle patches for combating infections. *Advanced Materials,* 2020. **32**(33): p. 2002129.

109. Wang, C.y., et al., Advances in antimicrobial organic and inorganic nanocompounds in biomedicine. *Advanced Therapeutics,* 2020. **8**: p. 2000024.

110. Umut, E., Surface modification of nanoparticles used in biomedical applications. *Modern Surface Engineering Treatments,* 2013. **20**: pp. 185–208.

111. Stöber, W., A. Fink, and E. Bohn, Controlled growth of monodisperse silica spheres in the micron size range. *Journal of Colloid and Interface Science,* 1968. **26**(1): pp. 62–69.

112. Deng, Y.-H., et al., Investigation of formation of silica-coated magnetite nanoparticles via sol–gel approach. *Colloids and Surfaces A: Physicochemical and Engineering Aspects,* 2005. **262**(1–3): pp. 87–93.

113. Santra, S., et al., Synthesis and characterization of silica-coated iron oxide nanoparticles in microemulsion: the effect of nonionic surfactants. *Langmuir,* 2001. **17**(10): pp. 2900–2906.

114. Yi, D.K., et al., Nanoparticle architectures templated by SiO_2/Fe_2O_3 nanocomposites. *Chemistry of Materials,* 2006. **18**(3): pp. 614–619.

115. Yoon, T.J., et al., Specific targeting, cell sorting, and bioimaging with smart magnetic silica core–shell nanomaterials. *Small,* 2006. **2**(2): pp. 209–215.

116. Yi, D.K., et al., Silica-coated nanocomposites of magnetic nanoparticles and quantum dots. *Journal of the American Chemical Society,* 2005. **127**(14): pp. 4990–4991.

117. Liao, H., C.L. Nehl, and J.H. Hafner, Biomedical applications of plasmon resonant metal nanoparticles. 2006.

118. Krumdieck, S.P., et al., Nanostructured TiO_2 anatase-rutile-carbon solid coating with visible light antimicrobial activity. *Scientific Reports,* 2019. **9**(1): pp. 1–11.

119. Wong, M.-S., et al., Visible-light-induced bactericidal activity of a nitrogen-doped titanium photocatalyst against human pathogens. *Applied and Environmental Microbiology,* 2006. **72**(9): pp. 6111–6116.

120. Kim, S., et al., Drug-loaded titanium dioxide nanoparticle coated with tumor targeting polymer as a sonodynamic chemotherapeutic agent for anti-cancer therapy. *Nanomedicine: Nanotechnology, Biology and Medicine,* 2020. **24**: p. 102110.

121. Kim, M.H., et al., Environmental disinfection with photocatalyst as an adjunctive measure to control transmission of methicillin-resistant *Staphylococcus aureus*: a prospective cohort study in a high-incidence setting. *BMC Infectious Diseases,* 2018. **18**(1): p. 610.

122. Mathew, S., et al., Cu-doped TiO_2: visible light assisted photocatalytic antimicrobial activity. *Applied Sciences,* 2018. **8**(11): p. 2067.

123. Reid, M., et al., How does a photocatalytic antimicrobial coating affect environmental bioburden in hospitals? 2018.

124. Nakano, R., et al., Photocatalytic inactivation of influenza virus by titanium dioxide thin film. *Photochemical & Photobiological Sciences,* 2012. **11**(8): pp. 1293–1298.

125. Yamaguchi, K., et al., A novel CD4-conjugated ultraviolet light-activated photocatalyst inactivates HIV-1 and SIV efficiently. *Journal of Medical Virology,* 2008. **80**(8): pp. 1322–1331.

126. Lee, J., K. Zoh, and G. Ko, Inactivation and UV disinfection of murine norovirus with TiO_2 under various environmental conditions. *Applied and Environmental Microbiology,* 2008. **74**(7): pp. 2111–2117.

127. Häggström, J., et al., Virucidal properties of metal oxide nanoparticles and their halogen adducts. *Nanoscale,* 2010. **2**(4): pp. 529–534.

128. Siddiqi, K.S., A. ur Rahman, and A. Husen, Properties of zinc oxide nanoparticles and their activity against microbes. *Nanoscale Research Letters,* 2018. **13**(1): pp. 1–13.

129. Tavakoli, A., et al., Polyethylene glycol-coated zinc oxide nanoparticle: an efficient nanoweapon to fight against herpes simplex virus type 1. *Nanomedicine,* 2018. **13**(21): pp. 2675–2690.

130. Sirelkhatim, A., et al., Review on zinc oxide nanoparticles: antibacterial activity and toxicity mechanism. *Nano-Micro Letters*, 2015. **7**(3): pp. 219–242.
131. Hu, X., et al., In vitro evaluation of cytotoxicity of engineered metal oxide nanoparticles. *Science of the Total Environment*, 2009. **407**(8): pp. 3070–3072.
132. Abdelmohsen, A. and N. Ismail, Morphology transition of ZnO and Cu_2O nanoparticles to 1D, 2D, and 3D nanostructures: hypothesis for engineering of micro and nanostructures (HEMNS). *Journal of Sol-Gel Science and Technology*, 2019: pp. 1–16.
133. Zhang, Y., et al., Synthesis, characterization, and applications of ZnO nanowires. *Journal of Nanomaterials*, 2012. **2012**: pp. 1–22.
134. Padmavathy, N. and R. Vijayaraghavan, Enhanced bioactivity of ZnO nanoparticles—an antimicrobial study. *Science and Technology of Advanced Materials*, 2008. **9**(3): p. 035004.
135. Stoyanova, A., et al., Synthesis and antibacterial activity of TiO_2/ZnO nanocomposites prepared via nonhydrolytic route. *Journal of Chemical Technology and Metallurgy*, 2013. **48**(2): pp. 154–161.
136. Jašková, V., L. Hochmannová, and J. Vytřasová, TiO_2 and ZnO nanoparticles in photocatalytic and hygienic coatings. *International Journal of Photoenergy*, 2013. **2013**: pp. 1–7.
137. Hochmannova, L. and J. Vytrasova, Photocatalytic and antimicrobial effects of interior paints. *Progress in Organic Coatings*, 2010. **67**(1): pp. 1–5.
138. Chung, C.-J., et al., Photocatalytic TiO_2 on copper alloy for antimicrobial purposes. *Applied Catalysis B: Environmental*, 2008. **85**(1–2): pp. 103–108.
139. Heinlaan, M., et al., Toxicity of nanosized and bulk ZnO, CuO and TiO_2 to bacteria Vibrio fischeri and crustaceans *Daphnia magna* and *Thamnocephalus platyurus*. *Chemosphere*, 2008. **71**(7): pp. 1308–1316.
140. Shen, L., H. Zhang, and S. Guo, Control on the morphologies of tetrapod ZnO nanocrystals. *Materials Chemistry and Physics*, 2009. **114**(2–3): pp. 580–583.
141. Jang, J.S., et al., Topotactic synthesis of mesoporous ZnS and ZnO nanoplates and their photocatalytic activity. *Journal of Catalysis*, 2008. **254**(1): pp. 144–155.
142. Mahmud, S., et al., Nanostructure of ZnO fabricated via French process and its correlation to electrical properties of semiconducting varistors. *Synthesis and Reactivity in Inorganic and Metal-Organic and Nano-Metal Chemistry*, 2006. **36**(-2): pp. 155–159.
143. Kakiuchi, K., et al., Fabrication of mesoporous ZnO nanosheets from precursor templates grown in aqueous solutions. *Journal of Sol-Gel Science and Technology*, 2006. **39**(1): pp. 63–72.
144. Mahmud, S. and M.J. Abdullah. Nanotripods of zinc oxide. in 2006 *IEEE Conference on Emerging Technologies-Nanoelectronics*, 2006, IEEE, pp. 1–7.
145. Ding, Y. and Z.L. Wang, Structures of planar defects in ZnO nanobelts and nanowires. *Micron*, 2009. **40**(3): pp. 335–342.
146. Xie, J., et al., Morphology control of ZnO particles via aqueous solution route at low temperature. *Materials Chemistry and Physics*, 2009. **114**(2–3): pp. 943–947.
147. Soares, N., et al., Active and intelligent packaging for milk and milk products, 2016.
148. Baum, M.K., G. Shor-Posner, and A. Campa, Zinc status in human immunodeficiency virus infection. *The Journal of Nutrition*, 2000. **130**(5): pp. 1421S–1423S.
149. Hiller, J.M. and A. Perlmutter, Effect of zinc on viral-host interactions in a rainbow trout cell line, RTG-2. *Water Research*, 1971. **5**(9): pp. 703–710.
150. Restuccia, D., et al., New EU regulation aspects and global market of active and intelligent packaging for food industry applications. *Food Control*, 2010. **21**(11): pp. 1425–1435.
151. Espitia, P.J.P., et al., Zinc oxide nanoparticles: synthesis, antimicrobial activity and food packaging applications. *Food and Bioprocess Technology*, 2012. **5**(5): pp. 1447–1464.

152. Jalal, R., et al., ZnO nanofluids: green synthesis, characterization, and antibacterial activity. *Materials Chemistry and Physics*, 2010. **121**(1–2): pp. 198–201.

153. Emami-Karvani, Z. and P. Chehrazi, Antibacterial activity of ZnO nanoparticle on gram-positive and gram-negative bacteria. *African Journal of Microbiology Research*, 2011. **5**(12): pp. 1368–1373.

154. Seil, J.T. and T.J. Webster, Antimicrobial applications of nanotechnology: methods and literature. *International Journal of Nanomedicine*, 2012. **7**: p. 2767.

155. Colon, G., B.C. Ward, and T.J. Webster, Increased osteoblast and decreased Staphylococcus epidermidis functions on nanophase ZnO and TiO$_2$. *Journal of Biomedical Materials Research Part A: An Official Journal of The Society for Biomaterials, The Japanese Society for Biomaterials, and The Australian Society for Biomaterials and the Korean Society for Biomaterials*, 2006. **78**(3): pp. 595–604.

156. Seil, J.T., E.N. Taylor, and T.J. Webster. Reduced activity of Staphylococcus epidermidis in the presence of sonicated piezoelectric zinc oxide nanoparticles. in 2009 *IEEE 35th Annual Northeast Bioengineering Conference*, 2009, IEEE, pp. 1–22.

10 Nanomaterials
Environmental Health and Safety Considerations

10.1 INTRODUCTION

Existing in the transitional zone between molecule and atom, nano-sized materials possess unique and peculiar properties unlike their macro-sized equivalents. Employing novel techniques allows for the fabrication of materials with improved conductivity, magnetism, optical sensitivity, mechanical strength, and performance as catalysing agents including biomedical agents, sensors, and electronic devices. Capitalizing on these properties has permitted nanotechnology to undergo a veritable revolution, enhancing and improving a wide array of products, industries, and services – in particular, the construction sector. However, the establishment of informative guidelines and regulations that both permit nanomaterials to be explored, while simultaneously providing sufficient protection for the environment, is still crucial, and a comprehensive assessment must be performed to evaluate their potential effects.

Employed within the construction industry for innovation, nanocomposites and nanomaterials may have unique and remarkable physical and chemical properties. However, it is not truly known for certain what potential unintended consequences may arise from their use that have an impact on human health and the environment at large. Given these materials are likely going to continue to be used, it is essential to explore the high-probability risk scenarios for some of the most commonly utilized nanomaterials, such as carbon nanotubes (CNTs), TiO_2, and quantum dots (QDs). The adverse toxicological and biological effects corresponding to these nanomaterials, along with their modes of action, are reviewed to supply a risk perspective and relative comparison. In order to inform future responsible manufacturing companies, about the use and disposal of nanoparticles in the construction contexts, this article concludes by identifying critical knowledge gaps and outlines the additional research required. This is carried out in alignment with the ongoing multidisciplinary action of nanomaterial risk assessments relative to the environment.

Nanomaterials can considerably modify the properties of construction materials and even improve their performance. Despite their merits, considerable research has reported that nanomaterials pose a potential risk to human health. For this reason, it is important to fully comprehend the effects of nanomaterials on human health and the environment throughout all phases of their life cycle, including manufacturing, construction use, and recycling, in order to ensure their responsible usage.

DOI: 10.1201/9781003196143-10

10.2 SAFETY CONSIDERATIONS

Nanomaterials are so many in number and so varied, it is safe to assume that massive quantities of these materials will eventually be produced. Moreover, the introduction of entirely new nanomaterials triggers both the requirement of adequate risk assessment procedures and suitable communication measures surrounding those risks. Presently, new nanomaterials are analysed in a manner similar to that used for chemicals, food, and consumer products, which is unsurprisingly both inefficient and insufficient. The challenges presented when characterizing nanomaterials and generating a standardized processing approach become substantial bottlenecks to the process. However, there do exist several techniques – laser ablation inductively coupled plasma mass spectrometry – that could stand to meet the needs of such processes.

This chapter sets out a suggested approach to reliably describe, characterize, and quantify the degree of human exposure to nanomaterials resulting from food packaging. Results acquired from such an approach are analysed, discussed, and concluding key criteria considered within the context of material development and regulation (criteria include solubility and fibre rigidity). Despite illustrating differing results dependent upon the application field, an investigation of public opinion in Germany found a noticeably positive perception of nanotechnology. This result also showed that opinion could be improved through the assurance of safety; accurate end-product control and toxicological testing are the key areas that ensure improvement of material characterization.

Categorized from safe to unsafe (as defined by the nanomaterials' release status detailed below), the interaction and exposure for the construction workers' biological systems and the nanomaterials are classified and described [1–4]. Nonreleasable nanomaterials (NRNs), such as the engineered materials in fire-resistant suits [5], do not interact with the environment nor with the workers under normal and standard working conditions. By contrast, releasable nanomaterials (RNs) do engage with the construction environment and workers' biological systems existing in an unsafe state, and readily releasing and interacting with anything in proximity.

Despite existing awareness surrounding the potential risks for working with construction nanomaterials, and the notion that these materials may even pose risks to end users, hazard information remains limited [7]. Consequently, the Occupational Safety and Health Administration (OSHA) has no recourse to mitigate the unknown hazards of NRs, as it is without regulations nor enforceable exposure limits; this is regardless of the fact that nanomaterial contamination can take place at any time during the manufacturing, packaging, and transport of construction materials, their use on-site during construction, and after the work is complete during the operational phase. For example, a number of workers were shown to have been exposed to more than the recommended limit of titanium dioxide during the packaging process in a study conducted by Al-Bayati and Al-Zubaidi [8]. In a recent move to promote safe working practices, the CPWR – The Center for Construction Research and Training developed a toolbox talk strongly recommending and endorsing the use of high-efficiency particulate air (HEPA) filters when handling nanomaterials. This was in response to the discovery that construction nanomaterials can be converted into unintended forms when mass-manufactured [9], such as carbon-based nanomaterials becoming airborne when prepared as a solution. However, it is worth noting that HEPA filters were never designed – to capture particles of under 300 nm in size, making it unlikely to eliminate the hazard, even though they may still serve to mitigate it.

Nanomaterials are considered as emerging environmental contaminants. Their origin can be natural [10], incidental [11], or from manufacturing processes [12]. Incidental nanomaterials are those generated as side products of anthropogenic processes [13,14], whereas manufactured nanomaterials (MNMs) are deliberately produced with specific properties [15]. Exposure to both types is currently being investigated and these may enter air, water, and soil media from a range of routes. Physicochemical and biological transformations make nanomaterials highly reactive in both environmental and biological systems, which may alter their fate, dispersion, and toxicity compared with their larger counterparts [1,16,17].

Development and market introduction of new nanomaterials trigger the need for an adequate risk assessment of such products alongside suitable risk communication measures. Current application of classical and new nanomaterials is analysed in the context of regulatory requirements and standardization for chemicals, food, and consumer products. The challenges of nanomaterial characterization as the main bottleneck of risk assessment and regulation are presented. In some areas, for example, quantification of nanomaterials within complex matrices, the establishment and adaptation of analytical techniques such as laser ablation inductively coupled plasma mass spectrometry and others are potentially suited to meet the requirements. As an example, we provide here an approach for the reliable characterization of human exposure to nanomaterials resulting from food packaging. Furthermore, results of nanomaterial toxicity and ecotoxicity testing are discussed, with concluding key criteria such as solubility and fibre rigidity as important parameters to be considered in material development and regulation. Although an analysis of the public opinion has revealed a distinguished rating depending on the particular field of application, a rather positive perception of nanotechnology could be ascertained for the German public in general. An improvement of material characterization in both toxicological testing as well as end-product control was concluded as being the main obstacle to ensure not only safe use of materials but also wide acceptance of this and any novel technology in the general public.

The interaction between nanomaterials and construction workers' biological systems can be classified in a range from safe to unsafe, as defined by the nanomaterials' release status [8,18,19]. Under normal working conditions, safe nanomaterials do not release or interact with workers or the construction environment [i.e. NRNs], for example, tissue-engineered materials in fire resistance suits [5, 6]. On the other hand, unsafe nanomaterials do release and interact with workers or the construction environment [i.e. RNs].

Exposure to nanomaterials can occur during the manufacturing and packaging of construction materials, the construction phase, or the operation phase. For example, a study conducted by Al-Bayati and Al-Zubaidi [8] showed that workers' exposure to titanium dioxide in the packaging process exceeded the recommended exposure limit (REL). As another example, construction nanomaterials can be transformed into new forms when manufactured on a large scale [9]. Additionally, they found that carbon-based nanomaterials can be converted into airborne nanomaterials when prepared as a solution. However, while characteristics of construction nanomaterials may increase toxicity hazards for construction workers during construction phases as well as for final users, hazard information is still limited [7]. As a result, there are no regulations or enforceable exposure limits adopted by the OSHA to mitigate the potential hazards of NRs. Recently, CPWR produced a toolbox talk that recommends the use of HEPA filters for construction workers who are handling nanomaterials. However, it must be clear that HEPA filters are not designed to capture nanoparticles since they are efficient in removing particles of 300 nm (i.e. 0.3 μm) or more in diameter. Therefore, HEPA filters may mitigate the hazard but not eliminate it.

10.3 POTENTIAL RISKS AND CONCERNS

As mentioned before, a significant impact stands to be made by the use of nanotechnology within the construction industry, not only from a perspective of enhancing material properties but also because a high proportion of all energy used by the world is consumed by commercial and residential buildings, in their lighting, heating, and air conditioning. Overtaken thus far by the adoption of nanotechnology within fields such as biomedical and electronics, the construction industry has been making up lost ground in their pursuit of innovation using a variety of MNMs in recent years. However, as alluded to previously, adoption of novel technologies does not come without risks; the potential dangers to the environment and human health posed by nanomaterials should not go unconsidered. This is true even if the goal in their use is to preserve the environment, by utilizing the energy-conserving functions provided by nanomaterials, their full lifespan must still be contemplated, as highlighted in a recent review by Rice University scientists. Unintended consequences could be far severe than those it was intended to prevent. Furthermore, the authors indicate that MNMs, especially CNTs, can be accidentally or incidentally introduced to the environment at various stages of their life cycle.

Within their work at Rice, they go on to detail the importance of a holistic MNMs' life cycle exposure profiling approach, stipulating without that level of meticulousness, critical impacts on ecosystem and human health cannot be avoided. They maintain that, as a result of no regulation being presently in place despite growing concerns, a number of MNMs should be regarded as 'potential emerging pollutants' until contradicting information surfaces, as there are many related risks to environmental and public health that are being disregarded without that regulation. Furthermore, they describe the element of unpredictability of the natural environment; once distributed into it, MNMs may transform in diverse chemical, biological, and physical fashions, altering their properties, effects, and ultimate fate.

The potential routes along which MNMs can be released into the environment are many and often. From occupational exposure, when the material is first being prepared, during any coating, moulding, incorporating, or compounding to contamination during installation, construction, maintenance, repair, renovation. Finally, to decommissioning or demolition processes. Even beyond this stage, further risks arise when solid nanomaterials reach landfills or get disposed of in incinerators. Delivery methods and approaches affect these risks, also: aerosolization of MNMs, adhesive wear, abrasion and corrosion, and manufacturing process wastewater effluent outlets all have additional risks, specific to the method and altering the resultant hazard.

10.4   RISK OF NANOMATERIALS

In the absence of more specific information, industrial hygienists and construction workers alike are best served operating under the assumption that any risks posed by nanoobjects – a catch-all term to cover nanofibres, nanoparticles, and nanoplates – are distinctly greater than those of the macro-material. This is as the size of nanoparticles predisposes them to greater mobility inside food chains and living systems, in addition to greater bioreactivity. Even though a study published in 2009 in the *Journal of Nanoparticles and Nanotechnology* [9] shared the discovery that a number of nanoparticles are capable of infiltrating across the blood–brain barrier, as yet the level of public pressure to avert these risks is absent. Given the rapidly growing quantity of nanomaterials, this area of research – the information from which may find application within other occupational exposure limit calculations – has found itself under particular observation by a large number of researchers.

A number of established journals have published findings already associating pathogenicity to chronic inflammatory processes for tubular fibres, like CNTs, and low-toxicity bio-persistent dusts, such as $nTiO_2$. In 2008 the *Journal of Toxicological Sciences and Nature Nanotechnology* illustrated a relationship between asbestos-like pathogenicity in mice, including mesothelioma, and CNTs; while more recently, a 2014 study in the *American Journal of Physiology* edition on Lung Cellular and Molecular Physiology detailed that asbestos fibres and CNTs were similarly still detectable in mice, causing a number of adverse health effects and more than 1-year post-exposure. It is perhaps unsurprising, then, that given the International Agency for Research on Cancer (IARC) and Occupational Safety and Health (NIOSH) agree in their categorization of nanoscale TiO_2 as a potential inhalation occupational carcinogen, the NIOSH has since issued RELs for both CNTs and ultrafine titanium dioxide (TiO_2). Subsequently, and without any other options to base their decision from, the British Standards Institute (BSI) proposed that the 2007 NIOSH REL for TiO_2 may serve as a reasonable starting point on which to judge the relative hazard of other similar insoluble nanoparticles.

When designing materials and performing their risk assessments, it is important to consider its factors and how they might influence its toxic potential. In 2006, an article identified a number of these factors, including its shape, solubility, degree of aggregation, chemical composition, surface charge, and surface structure. It is possible to adjust the design of MNMs and other nanoparticles to make them safer, once the high-risk avenues are identified; for example, the capping of a CNT with a single C_{60} fullerene effectively prevents the new composite from being able to penetrate the pleura.

Even though many of these studies raising concern points have existed in the public domain for almost a decade already, it appears that the long-term human and ecological health effects of engineered nanoparticle (ENP) exposure are still yet to be understood. A cause for concern, as despite this being the case, the use and deployment of MNMs has not slowed. In the construction industry, these knowledge gaps break even wider, particularly falling on the approaches for risk assessment. They are still functioning of unknown, unmonitored, and under-explored potential exposures to MNMs, it is essentially impossible for them to be anything more useful. Furthermore, even laboratory research and experimentation is dissuaded by the high risk of fire or explosion resulting from highly combustible nanoparticle powders. As long as the epidemiological data continue to be lacking, there remains no other source to rely upon for this crucial and overdue information.

10.5 TOXICITY OF NANOMATERIALS

As is common with emerging technologies, the potential benefits of the use of a new product, material, or process are frequently obvious, whereas the assessment of risks requires the investment of time. This poses a troubling dilemma for industrial hygienists in particular, as by the time they have gathered the information required to make informed decisions on potential hazards and risk mitigation, a novel technology can already be widely prevalent, extensively used, and well established. Unfortunately, this is also the case for nanotechnology. Since a present and active part of numerous industries, rapidly asserting itself further within the field of construction, this leaves very little time to act in any meaningful preventative manner should issues be discovered. While the allure of the considerable adjustments to construction material properties through use of nanotechnology is significant – developments in this area can lead to reduced resource consumption, improved energy efficiency, and biodegradability – it must not be permitted to deprioritize considerations of the unknown occupational risks these new materials could be bringing with them.

Despite appearing permanently embedded within building materials, or fixed in their other construction applications, many MNMs can still cause cellular toxicity through various mechanisms (Figure 10.1). The most significant of which include the

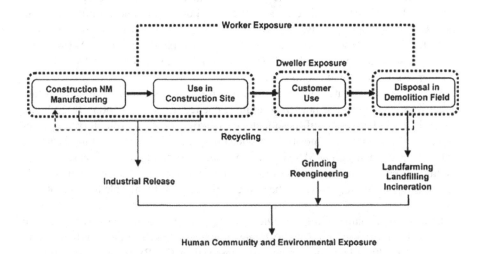

FIGURE 10.1 Possible exposure scenarios during the life cycle of MNMs used in construction [20].

generation of reactive oxygen species (ROS) that inflict oxidative stress, such as can be caused by TiO_2; the release of toxic heavy metals and other components, occurring with QDs; the direct oxidation through contact with cell constituents, nC_{60}, for example; disruption of cell wall integrity, as in SWNTs; and nucleic acid damage. As the nanomaterials most frequently used, CNTs and $nTiO_2$ are the ones with the greatest quantity of available literature and research into possible toxic effects they can incite. This is discussed at greater length below.

The photoreactive properties of the nanomaterial $nTiO_2$ that make it useful, for antimicrobial activity and self-cleaning surface coatings for example, are also what make it dangerous. With or without UVA radiation to trigger ROS production, to mammalian cells, $nTiO_2$ presence produces inflammation, cytotoxicity, and DNA damage [22]. Depending upon its relative morphology, $nTiO_2$ can achieve mobility inside cells and through cell membranes, in addition to potentially triggering the signalling process for ROS synthesis upon encounters and interactions with phagocytic cells. In terms of the aforementioned antimicrobial properties, $nTiO_2$ has demonstrated a capacity to inhibit *Micrococcus luteus*, *Aspergillus niger*, *Bacillus subtilis*, and *Escherichia coli* in solar disinfection surface coatings [27].

In addition to the negative effects previously described, CNTs have also been found to exert pulmonary toxicity in a number of mammals [16]. The mechanism by which this happens is same way CNTs are able to exhibit antibacterial functions: through oxidative stress, which then causes cell wall damage, or through direct physical interaction. A particular allotrope of carbon, buckminsterfullerene (C_{60}), is water insoluble [24], though with its agglomeration using transitional solvents or patiently sustained stirring over the long term imparts a degree of water stability, influencing its potential exposure and toxicity as a consequence [26]. This leads to C_{60} suspensions that are relatively water-stable; this is what is referred to as nC_{60} [18], the nanoparticle form of buckminsterfullerene, and in this format, it exhibits broad-spectrum antibacterial behaviours. Despite early studies attributing the mechanism of nC_{60} cytotoxicity in eukaryotic systems to oxidative stress, induced from the ROS production, it has recently been demonstrated that detectible levels of ROS are not present [26], and instead the antibacterial properties are enabled through the direct oxidation of the cell [17]. It is worth noting, however, that this toxicity is readily influenced and mitigated through introduction of dissolved natural organic matter to the solvent, as they coat the particle effectively and decrease their availability to react.

10.6   USING NANOMATERIALS SAFELY

The question of how to utilize nanomaterials safely does not find itself wholly resolved, even though it is clear that discovering it is crucial to improving the performance of infrastructure and buildings. These nanoscale fibres and particles could be already contributing to a problem that the scientific community is as yet completely unaware of, or in the ways of which we know that they can. Thin strands carried airborne can acquire behaviour patterns akin to asbestos. Limited information is available for workers and manufacturers alike on keeping safe while handling these materials, while it is commonplace to appreciate the necessity of greater regulation.

Given that estimates place up to half of all new building materials in 2025 as containing nanomaterials, this information is urgently sought. This was the motivation for the research team at Loughborough University, when they investigated where these materials are used, to what extent, a number of potential risks, and how might the workers on the 'front line' mitigate these risks. It was funded in part by the Institution of Occupational Safety and Health (IOSH), in order to produce a framework and a measure of guidance.

An additional challenge facing the generation of a set of guidelines as such, or indeed, any other form of regulation, is that the way health and safety legislation is applied in varying countries. It may not be mandatory for manufacturers to specify information about the type of nanomaterial, or the approach with which it was used, resulting in largely unreliable and inconsistent labelling systems.

10.7 MITIGATION OF PUBLIC AND ENVIRONMENTAL HEALTH IMPACTS

According to the team of researchers at Loughborough, it is theoretically possible to re-engineer MNMs to adjust their designs to make them safer, greener, while retaining their useful properties. However, recent examples exist to show for the success of such an approach. The substitution of branched alkyl benzene sulfonate detergents with biodegradable linear homologues had an excellent effect on reducing the excessive foaming; and the employment of safer, less persistent hydrochlorofluorocarbons in order to replace and displace ozone-depleting chlorofluorocarbons.

One such approach to improving the safety of MNMs recommended by the team involves first discerning which of the structures and associated properties are the ones that make it harmful, and prioritize those by which receptor is at highest risk. Given that over-intensive detoxification can serve to invalidate the product, by rendering it no longer able to perform the function it was originally intended for, focusing alternatively upon exposure control appears to be recommended. As opposed to suppressing the intrinsic reactivity as a whole, this frequently can provide a more targeted approach to alleviate toxicity risks without taking the chance on material failure.

Re-emphasizing the potential for energy conservation in the application of MNMs within the construction industry, the authors made a number of notable observations in their concluding remarks:

"Opportunities for energy savings – other than using manufactured nanomaterials to harvest solar or other forms of renewable energy – include improved thermal management by using silica nanoparticles in insulating ceramics and paint/coating that enable energy conservation and solar-powered self-cleaning nano-TiO_2-coated surfaces. Additional opportunities include the use of quantum dots and carbon nanotubes to improve the efficiency of energy transmission, lighting, and/or heating devices, as well as incorporation of fullerenes and graphene to enhance energy storage systems such as batteries and capacitors that harvest energy from intermittent, renewable sources (e.g., solar and wind)".

Additionally, the indirect contribution towards the reduction of wasted energy sourced from designing and fabricating nanomaterials that extend the durability of structures is significantly more than might be intuitive, as they reduce the requirement for the input of energy that would be otherwise required to repair or replace infrastructure that has deteriorated over its lifespan. Another benefit of MNMs can be found in the active replacement of extremely harmful environmental pollutants, including mercury and lead, paving the way for a greener construction industry.

10.8   RISK COMMUNICATION AND TECHNOLOGICAL IMPACT ASSESSMENT

The initial enthusiasm surrounding nanotechnology as a crucial enabler in the 21st century led to some exceedingly enthusiastic early forecasts, predicting market volumes between 1 and 3 trillion USD [21] between 2000 and 2010. While some of the early excitement appears to have receded, this frequently is a result of companies shifting attention towards the optimal solutions from function and cost perspectives, oftentimes holding the launch of novel products back.

Throughout the globe, the applications most customarily used for nanotechnology within the consumer product sphere are in the categories of home and fitness, home and garden, and automotive; these are closely followed by food and beverage production and then coatings [22]. As mentioned previously, nano-silver is one of the most regularly used nanomaterials, with its primary function as an antimicrobial agent, it can be found in 25% of all nano-enabled products currently in market.

While it is widely accepted that the development of nanotechnology appears to be following a double-boom pattern with its publications and patents, this is not typical of a technology cycle. Normally, an acceleration before the second boom is preceded by a stagnation of scientific trends in advance of the first boom. It is also commonplace that such activities be more stable and fluctuate less in their progress, particularly given companies and businesses tend to respond more quickly to unrealized expected results. In a standard format, the cycle usually lasts 15 years or more between the first and second booms. In this scenario, no substantial markets are in place in anticipation for the second boom. Provided no extreme changes are seen, the current phase of consumer orientation, enterprise, and research is set to inspire a wide and diverse array of new product properties and opportunities over the coming decades.

There are a number of contributing factors whether the disclosure of nanotechnology on product descriptions is advantageous or disadvantageous, and it predominantly relies upon the respective stakeholders. In some large companies, questions have been raised and concerns aired regarding nanospecific labelling and information sharing, because they will endure stigmatization and financial burden from the requirement to do so. Contrastingly, small-to-medium enterprises (SMEs) particularly in the paint industry have a need for additional information on the NMs in use. From an additional perspective, non-governmental institutions expect a certain degree of transparency. Lastly, it was a point of concern that consumers could simply forget what nanotechnology is and may not facilitate its spread and differentiation into other products.

In order to continue pursuing the development of nanotechnology, a consideration needs to be made to include – on a broad level and in a continuous fashion – all involved stakeholders and their perspectives. Appropriate risk management techniques should be employed, dependent upon whether free, bound, or embedded particles are the most likely threat, in a process moderated by neutral parties. Such a risk assessment was carried out by the NanoView project, studying the perception and public and media opinion regarding nanotechnology, finding that while 41% of Europeans are in a positive state of mind about nanotechnology, 40% are still undecided (out of the EU 27 member states in 2010). This average awareness differs greatly at even a country-to-country level; with positive attitude at 76% in Switzerland, 65% in Germany and 62% in the United States, whereas the average for all of the European Union member states drops to 46%. It was observed that a link existed between desire to purchase nano-enabled products and the expected intensity of human exposure, perhaps explaining why then that the surface coating and care products are the easiest to approve of, with an acceptance rating superior to 70%. This drops to 60% for textiles, 30% for cosmetics, to the lowest value of 20% when it comes to food.

In an analysis of the public perception of people living in Germany and media coverage towards nanotechnology, the periods of 2000–2007 and 2008–2012 were included for a gathering of total number of news media articles for the given subject. The total quantity of media articles (regardless of size, length, or quality) was around 1696 in the first period, declining to 591 in the second; however, the opposite phenomena was observed in the percentage of articles placed in the science sections of newspapers and news magazines, with 58.5% in the first period and 66.5% in the second [21]. It is worth noting however that the topic of nanotechnology is not frequently raised in Germany at present – not because it is taboo, but instead because it is simply a highly specialized subject that not many are keen to involve themselves into.

10.9   ENVIRONMENTAL IMPLICATIONS

With the rise of nanoparticle applications and demands from previously unrelated industries, the exposure and production of nanoparticles have risen. Thus, through transport, erosion, washing, disposal, and erosion of nanoparticle-enriched products, nanoparticles will find their way into ecosystems [23,24]. The impact of the introduction of nanoparticles into marine and land environments is currently ambiguous, with the direct results of nanoparticle exposure unclear. Materials that were previously deemed safe to environments must be reassessed at a nanoscale dimension due to the previously unforeseen interactions that nanoparticles experience [23]. Thus, it is pertinent to consider the implications of nanoparticles in the environment and judge the cost benefits of their use in industry. One of the key issues with nanoparticles is the way that interactions occur involving materials that were previously considered to be nontoxic. Specifically, although silver has hitherto been considered nontoxic, evidence suggests that exposure to silver nanoparticles in the embryotic stage of zebrafish can trigger development abnormality and/or death at concentrations at or above 0.19 nanomolar [23]. Furthermore, non-marine examples provide evidence of potential issues arising from environmental nanoparticle interactions, with silver nanoparticles disrupting the seed growth of a variety of plants, causing silver accumulation in shoots and bioaccumulation in green algae, rag worms, and gastropods [23].

Nanoparticles are released to the environment through a multitude of channels, both intentional and unintentional, via atmospheric emissions, and effluent, or through remediation of contaminated land or water [24,25]. Products enriched with nanoparticles comprise a proportion of the nanoparticles being deposited in the environment, and, thus, it may be postulated that as the use of nanoparticles increases, the same relationship will hold true for environmental nanoparticles [25]. ENPs are the primary concern of risk analysis regarding nanoparticles, but it is integral that incidental nanoparticles not be overlooked. Tyre rubber wear particles and brake pad wear particles are commonly neglected, despite contributing a significant portion of particulate emissions annually. A busy road with 25,000 cars travelling daily will generate up to 9 kg of particulate matter per kilometre of road, with no remediation [26]. This has led to the deposition of automotive-related particulate matter deposition in streams and lakes. A study of 149 sediment samples collected from watersheds in the United States, France, and Japan demonstrated a 97% detection rate, with values ranging from 14 to 5800 ppm in dry weight. An average value of 1000 ppm and mean of 440 ppm demonstrates a vast difference in concentration with France, on average, registering five times the concentration of particles than the United States or Japan. However, this is theorized to be due to the silt and clay content of the Seine River, France [27].

10.10   SUMMARY

Accompanied by standardization approaches and regulatory measures, high-volume nanomaterials (NMs) such as SiO_2 and carbon black play a major role for a variety of industrial applications. Despite still being in development phases for many applications, the appropriate analytical capacity for the characterization of materials and their properties are still necessary and fundamental; existing reports from toxicological inhalation studies already indicate steady increases of NM toxicity, as opposed to demonstrating entirely new nano-specific effects and outcomes. These effects are improbable, however, as a result of the rather arbitrary and flighty nature of the definition of 'nanomaterial'. Nevertheless, while nanodimensions themselves do not present toxicological hazards, the sheer quantity of new materials such as hybrids and composites makes further attention and study mandatory.

In light of the variety of NMs both in daily life and industrial processes alike, further advancement of analytical techniques is posed to be indispensable for three key reasons: first, the accurate quantitative description and characterization of exposure scenarios, such that ways to circumvent them might be explored; second, to facilitate the extended evaluation of potential adverse effects on humans; and third, the same consideration applied to environmental fates of nano-sized particulate matter, in their destiny, effects on the ecosystem, and ecotoxicity.

Elaborating the first point in particular, the procedures presently used for testing food and materials that come into contact with food can serve as a starting point from which to specialize and adjust the techniques for identifying appropriate human exposure levels through other products, such as cosmetics and textiles.

Development of the second point with the specific capabilities, such as membrane penetration and intracellular accumulation, of nanoparticles can be revealed through improved integration of biokinetic studies into existing toxicological testing schema.

When considering the potential impact of NM on ecotoxicity in the long term and on the environment, the identification of the differing transformations and alterations an NM can undergo during its different life cycle stages is vital, as these represent the routes by which said material's toxicological properties may unexpectedly change.

Although the integration and adoption of advanced analytical techniques, such as LA-ICP-MS, to biological matrices shows promise, it should be noted that it is significantly time-consuming and needs a much greater emphasis to be placed on nanosafety within those projects. Independently of whether NMs themselves have an intrinsic toxicity related to their size, a meticulous cataloguing of toxicological test systems would prove to enhance data quality and toxicokinetic understandings as a whole. As a crucial prerequisite for reliable risk assessments, this is also a necessity in the journey to gaining acceptance of novel technologies in the public space.

REFERENCES

1. Lowry, G.V., et al., Transformations of nanomaterials in the environment. *Environmental Science and Technology*, 2012. **46**: pp. 6893–6899, ACS Publications.
2. Klaine, S.J., et al., Nanomaterials in the environment: behavior, fate, bioavailability, and effects. *Environmental Toxicology and Chemistry: An International Journal*, 2008. **27**(9): pp. 1825–1851.
3. Singh, A.K., et al., Nanotechnology in road construction. in *AIP Conference Proceedings*, 2020, AIP Publishing LLC.
4. Samadi, M., et al., Influence of glass silica waste nano powder on the mechanical and microstructure properties of alkali-activated mortars. *Nanomaterials*, 2020. **10**(2): p. 324.
5. Alongi, J. and G. Malucelli, Cotton flame retardancy: state of the art and future perspectives. *RSC Advances*, 2015. **5**(31): pp. 24239–24263.
6. Meng, H., et al., A predictive toxicological paradigm for the safety assessment of nanomaterials. *ACS Nano*, 2009. **3**(7): pp. 1620–1627.
7. Krug, H.F., Nanosafety research—are we on the right track? *Angewandte Chemie International Edition*, 2014. **53**(46): pp. 12304–12319.
8. Al-Bayati, A.J. and H.A. Al-Zubaidi, Inventory of nanomaterials in construction products for safety and health. *Journal of Construction Engineering and Management*, 2018. **144**(9): p. 06018004.
9. Johnson, D.R., et al., Potential for occupational exposure to engineered carbon-based nanomaterials in environmental laboratory studies. *Environmental Health Perspectives*, 2010. **118**(1): pp. 49–54.
10. Kumar, P., et al., A review of the characteristics of nanoparticles in the urban atmosphere and the prospects for developing regulatory controls. *Atmospheric Environment*, 2010. **44**(39): pp. 5035–5052.
11. Kumar, P., P. Fennell, and A. Robins, Comparison of the behaviour of manufactured and other airborne nanoparticles and the consequences for prioritising research and regulation activities. *Journal of Nanoparticle Research*, 2010. **12**(5): pp. 1523–1530.

12. Peralta-Videa, J.R., et al., Nanomaterials and the environment: a review for the biennium 2008–2010. *Journal of Hazardous Materials*, 2011. **186**(1): pp. 1–15.
13. Azarmi, F., P. Kumar, and M. Mulheron, The exposure to coarse, fine and ultrafine particle emissions from concrete mixing, drilling and cutting activities. *Journal of Hazardous Materials*, 2014. **279**: pp. 268–279.
14. Kumar, P., et al., Nanoparticle emissions from 11 non-vehicle exhaust sources–a review. *Atmospheric Environment*, 2013. **67**: pp. 252–277.
15. Kumar, P., A. Kumar, and J.R. Lead, Nanoparticles in the Indian environment: known, unknowns and awareness. *Environmental Science and Technology*, 2012, **46**: pp. 7071–7072, ACS Publications.
16. Ju-Nam, Y. and J.R. Lead, Manufactured nanoparticles: an overview of their chemistry, interactions and potential environmental implications. *Science of the Total Environment*, 2008. **400**(1–3): pp. 396–414.
17. Khin, M.M., et al., A review on nanomaterials for environmental remediation. *Energy & Environmental Science*, 2012. **5**(8): pp. 8075–8109.
18. Kangas, H. and M. Pitkänen, Environmental, Health & Safety (EHS) aspects of cellulose nanomaterials (CN) and CN-based products. *Nordic Pulp & Paper Research Journal*, 2016. **31**(2): pp. 179–190.
19. Walker, G., R. Ball, and A. Gibb. Growth of nanomaterials in construction raises health and safety concerns. in *Proceedings of the Institution of Civil Engineers-Civil Engineering*, 2014, Thomas Telford Ltd.
20. Lee, J., S. Mahendra, and P.J. Alvarez, Nanomaterials in the construction industry: a review of their applications and environmental health and safety considerations. *ACS Nano*, 2010. **4**(7): pp. 3580–3590.
21. Laux, P., et al., Nanomaterials: certain aspects of application, risk assessment and risk communication. *Archives of Toxicology*, 2018. **92**(1): pp. 121–141.
22. Vance, M.E., et al., Nanotechnology in the real world: redeveloping the nanomaterial consumer products inventory. *Beilstein Journal of Nanotechnology*, 2015. **6**(1): pp. 1769–1780.
23. Yin, Y., et al., Source and pathway of silver nanoparticles to the environment, in Jingfu Liu and Guibin Jiang (eds) *Silver Nanoparticles in the Environment*, 2015, Springer: China, pp. 43–72.
24. Mohajerani, A., et al., Nanoparticles in construction materials and other applications, and implications of nanoparticle use. *Materials*, 2019. **12**(19): p. 3052.
25. Bhatt, I. and B.N. Tripathi, Interaction of engineered nanoparticles with various components of the environment and possible strategies for their risk assessment. *Chemosphere*, 2011. **82**(3): pp. 308–317.
26. Novotny, T.E., et al., Cigarettes butts and the case for an environmental policy on hazardous cigarette waste. *International Journal of Environmental Research and Public Health*, 2009. **6**(5): pp. 1691–1705.
27. Moerman, J. and G. Potts, Analysis of metals leached from smoked cigarette litter. *Tobacco Control*, 2011. **20**(Suppl 1): pp. i30–i35.

11 Sustainability and Environmental Benefits of Concrete with Nanomaterials

11.1 INTRODUCTION

Recently, with advances in knowledge made at a very rapid pace, the dissemination of this new knowledge has become the need of the hour. The concrete industry usually consumes a great part of the natural resources extracted from the planet, in which cement represents the greatest cause of this consumption. Therefore, a great deal of interest was oriented to replace the traditional ingredients of concrete structure by a new material to increase its durability and sustainability. Literature review shows that cement produced with nanomaterials results in much harder and cheaper cement that can improve the engineering and durable properties of modern concrete [1]. Currently, production of cement with nanomaterials as additives or replacement such as nano-silica, nanotitanium, nanoiron, and nanocarbon fibre has been recognized to enhance the concrete engineering properties [2–4]. Due to their properties at the ultra-fine level, they exhibit three main advantages;

i. Filling the small voids between the cement grains, acting as a filler effect, producing a denser microstructure. In addition, nanomaterials have the property of pozzolanic activity.
ii. Enhancing the hydration process at early age and formulate denser gel.
iii. Improving the bond strength between fine and coarse aggregates and cement pastes.

The current study in modern concrete indicates that smart cement has been perceived as an intelligent system which has different structural properties as compared to the traditional cement-based concrete. For instance, the self-cleaning, self-healing, and the self-sensing concrete have been associated with the capacity to respond to the environment through temperature, stress, or an external stimulus. The concept of "smartness" in the concrete has been achieved using the four main methods, namely material design composition, special processing, introducing other efficient components, and alteration of the microstructure. The selection of these methods has been tailored to achieve specific needs by improving the properties of the concrete, such as workability, strength, durability, ductility, serviceability, reliability, cost, among

DOI: 10.1201/9781003196143-11

other properties [5,6]. Over time, the construction of concrete production has changed to achieve desirable properties that are environmentally friendly, durable, and safe.

Advancement in nanotechnology has offered tools that can support the fabrication and development of materials which can enhance the functionality of the traditional cement-based materials. In a study by Jeevanandam et al. [7], it was demonstrated that the interest in utilization of nanomaterials in the concrete industry has been increased, especially the use of nanoparticles characterized by smart properties. Based on the previous studies and reports [8,9], it was noted that the cement industry accounted for approximately 6%–7% of carbon emissions while using a lot of energy during the production of concrete. From a traditional perspective, concrete is developed based on a set of prescribed specifications, which includes the degradation of a material over a particular duration. Mitigating the inevitable degradation calls for the use of a costly maintenance regime and in some cases, the use of a completely new approach [5,10].

With the development and growth of nanoscience and nanotechnology, the potential of nanomaterials to improve cement material properties increases and supports the sustainability [11]. There is a huge amount of agricultural, industrial, and construction waste materials around the globe such as fly ash, ceramic, palm oil fuel ash, and demolition concrete wastes. The concrete wastes estimated to 317 million tones (MT) in the United States, 510 MT in Europe, and 239 MT of waste in China during the period of 2009–2010 [12]. This level of wastage is unacceptable, especially with the depletion of the world's natural resources beyond the industrial revolution. This has led to a call for changes to the environmental management plans of construction sites, which now state that the materials on the site should be stockpiled and separated so that they may be reused and recycled for other construction projects. From these recycled products, various-sized aggregates using sieves can then be used in future projects to improve the sustainability of structures and reduce the overall embodied energy [12]. Furthermore, the construction industry is always looking at other ways to improve the environmental and economic sustainability. Even though using nanoparticles in concrete has improved constructability and resulted in greatly improved construction times, recycling these nanoparticles along with the health hazards associated with them for workers in demolition/construction and members of the communities in the area remains unanswered [13].

The recycling of concrete with nanoparticles has been investigated recently, and it has been found to possess higher compressive strength than that of typical recycled normal concrete [14]. One of the main concerns regarding the recycling of concrete without nanoparticles is that it is inferior to normally prepared concrete with regard to durability and mechanical properties. This is an issue when attempting to use recycled concrete for large-scale infrastructure and projects, which invite a higher risk, hence, may deter the selection of the recycled concrete. However, it has been found that the addition of nanoparticles to the recycled normal concrete can produce mechanical properties of the level of fresh normal concrete [14]. The strength and the microstructure are improved, but the workability of concrete is reduced with the addition of nanoparticles. It has also been determined that recycled concrete with nanoparticles can achieve comparable compressive strength to that of fresh normal concrete after 28 days, when the percentage by mass of nanosilica is 3% by mass [14]. The current research available is limited, and more work should be conducted to investigate the dynamic mechanical properties of the effects of nanoparticles with the recycled concrete and how this compares against fresh normal concrete in terms of impact loading [14].

11.2 LIFE CYCLE ASSESSMENT

Worldwide energy consumption has reached a level which was never observed before [15,16]. With the increase in population, the demand for construction materials has increased affecting energy consumption and carbon dioxide emission. To reduce energy demand and CO_2 emission, there is a dire need to produce not only sustainable construction materials but also develop other resources. In this section, the environmental data of various mortar components (fly ash (FA), ground blast furnace slag (GBFS), and nano glass powder) used to prepare 1 m^3 of alkali-activated mortars (AAMs) are calculated and presented in Table 11.1. The life cycle inventory of AAM production has been built according to life cycle of raw materials used as binder in the production process which included collection, transportation, crushing, grinding,

TABLE 11.1
Materials and Machines Information to Prepare 1 m^3 of AAMs

Item, Units	Amount
Truck speed, km/h	80
Diesel consumption, L/km	0.09
Diesel price, RM/L (RM: Malaysian Ringgit)	2.18
Truck volume, m^3	12
Transport charge of 1 m^3, RM/km	0.75
Glass density, kg/m^3	1280
GBFS density, kg/m^3	1860
FA density, kg/m^3	1350
Crushing machine power, Watt	435
Sieving machine power, Watt	250
Oven power, Watt	1200
Grinding machine power, Watt	750
Crushing machine capacity, m^3	0.08
Sieving machine capacity, m^3	0.05
Oven capacity, m^3	0.18
Grinding machine capacity, m^3	0.45
CO_2 emission for 1 L diesel, ton	0.0027
Energy consumption for 1 L diesel, GJ	0.0384
Electricity tariff, RM/kwh	0.516
CO_2 emission for 1 kwh electricity, tonne	0.00013
Energy consumption for 1 kwh electricity, GJ	0.0036

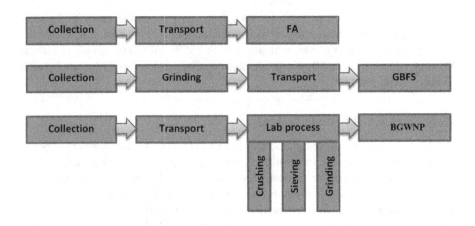

FIGURE 11.1 Life cycle of raw materials: FA, GBFS, and BGWNP.

and other treatments in the laboratory. The life cycle of raw materials presented in Figure 11.1 adopted as boundary in life cycle assessment. The data for fine aggregate, sodium hydroxide, and sodium silicate were not included in calculation as the content of these three materials is fixed for all AAM mixtures.

The total CO_2 emission, cost-effectiveness, and energy consumption of different alkali-activated binders were calculated to assess the mortar sustainability. Energy and cost requirements as well as the CO_2 emission of each material (Fly ash (FA), Ground blast furnace slag (GBFS), and Waste bottle glass nano powder (WBGNP)) were estimated depending on material life cycle including collection, transport, and lab treatments (such as crushing, sieving, oven drying, and grinding) as shown in Figure 11.1. As mentioned earlier, the FA was used in the lab as received, so the life cycle was included only for the collection and transport stages. GBFS was collected and ground before being transported from Ipoh to Johor (500 km distance). For BGWNP, the life cycle included collection, transport, and lab treatment such as cleaning, crushing, sieving, and grinding. In transportation stage and for all the raw materials, the truck engine capacity, volume, and speed were fixed to 0.09 L/km, 12 m³, and 80 km/h respectively. The local price of electricity, diesel, and transportation was considered; the calculation process is presented in Table 11.1. Meanwhile, the total CO_2 emission, cost, and energy consumption of each mixture depending on binder ratio (FA:GBFS:BGWNP) was calculated using respective Equations 11.1–11.4.

$$\text{Total CO}_2 \text{ emission} = \sum_{i=1}^{n} xi \left[(di \times fi \times zi) \right] \qquad (11.1)$$

where mi = mass of component i (ton per m³), di = transport distance (km), Di = diesel consumption (L/km), $k1i$ = CO_2 emission for 1 litre diesel, ton, Ei = total electricity consumption (kwh), and $k2i$ = CO_2 emission for 1 kwh electricity, tonne.

$$\text{Total energy consumption} = \sum_{i=1}^{n} xi \left[(di \times fi \times k2i) \right] \qquad (11.2)$$

where $k3i$ = energy consumption for 1 litre diesel, GJ, Ei = total electricity consumption (kwh), and $k4i$ = energy consumption for 1 kwh electricity, GJ.

$$\text{Total cost} = \sum_{i=1}^{n} xi \left[(di \times fi \times DPi) + Ti \right] \quad (11.3)$$

where DPi = diesel cost (RM/L), Ti = transport charge of $1\,m^3$ (RM/km), and EPi = electricity cost (RM per kwh).

$$\text{Electricity consumption of component } i\ (Ei) = \sum_{i=1}^{n} (MEi \times MPi) \quad (11.4)$$

where MEi = machine capacity (tonne per h) and MPi = machine power (kwh).

The effect of BGWNP replacing GBFS in AAMs on binder cost, carbon dioxide emission, and energy consumption is depicted in Figures 11.2–11.4, respectively. The AAM binder cost (Figure 11.2) decreased with the increase in BGWNP level. The cost dropped from RM 104.9 to 101.3, 97.7, 94.2, and RM 90.6 as the replacement level of GBFS by BGWNP increased from 0% to 5%, 10%, 15%, and 20%, respectively. In short, reuse of glass bottle waste in the form of nanopowder to replace GBFS contributed to save cost up to 13%.

Figure 11.3 shows the CO_2 emission of AAMs depending on the replacement level of GBFS by BGWNP. The CO_2 emissions of all AAMs binder were influenced by BGWNP content and their values dropped from 35.6 to 33.2, 30.8, 28.4, and 25.9 CO_2kg per tonne as the BGWNP content replacing GBFS raised from 0% to 5%, 10%, 15%, and 20%, respectively.

Figure 11.4 displays the influence of BGWNP on energy consumption of AAM binder. The energy consumption was directly proportional to the BGWNP content, whereby its increasing content from 0% to 5%, 10%, 15%, and 20% as GBFS replacement could reduce energy consumption from 0.551 to 0.544, 0.537, 0.527, and 0.523

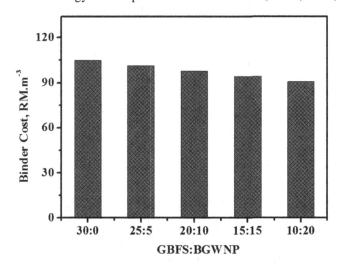

FIGURE 11.2 AAM binder cost with various ratios of GBFS to BGWNP.

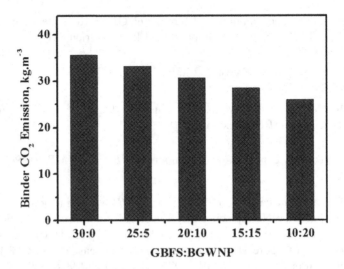

FIGURE 11.3   GBFS to BGWNP ratio-dependent CO_2 emission from AAMs.

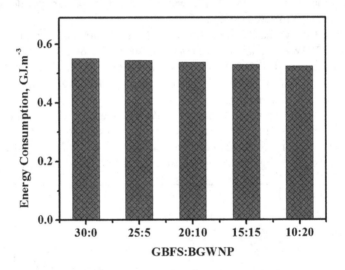

FIGURE 11.4   Influence of AAM binder energy consumption by BGWNP content.

Gl/t, respectively. Previous studies [17,18] also reported that a decrease in GBFS content in AAM mixtures reduced carbon dioxide, cost, and energy consumption.

11.3   SUSTAINABILITY OF NANOMATERIAL-BASED CONCRETE

Presently, nanotechnology is one of the most significant scientific and industrial breakthroughs of the 21st century. Nanomaterials offer great advantages towards concrete sustainability in construction fields such as energy storing, high performance, corrosion resistance, environmental remediation, and self-healing applications.

Sustainability is a celebrated topic nowadays, as nanomaterials become incorporated into concrete products in rising amounts, it may help to develop an understanding of their interaction with the environment. Nanotechnology has the potential to dramatically change the strength, sustainability, and whole properties of concrete. Hamers [19] have reported the broad range of complex nanomaterials required to understand the molecular-level design rules. Eventually, it is challenging to exploit the power of chemistry to guarantee that nanosystem-incorporated technologies can make better environmentally friendly products.

11.4   MERITS AND DEMERITS OF NANOMATERIALS FOR CONCRETE

In general, sustainable energy and the environment are one of the central priorities for researchers; triggering a massive capital investment on research to define new trajectories in construction materials' sustainability and consequent pollution abatement. This attempt by researchers to solve concrete problems by using nanomaterials as the modifying agent contributed to achieving many advantages. Moreover, deployment of nanomaterials in the self-healing, self-curing, self-cleaning concrete, and ultra-strength concrete are an emergent concept. High performance of nanomaterials affects positively towards the enhancement and development of concrete's performance and sustainability. The applied nanotechnology led to develop a molecular model for the hydration products (C–S–H gels) of cement [20] as shown in Figure 11.5. As well as enhancement of strength and sustainability of self-healing concrete using nanomaterials the controlled self-healing be better and the price of the material is considerably below epoxy-based materials.

The cost of self-healing concrete compared to conventional concrete is still high even with use of nanomaterials. Thus, self-healing concrete is a probable product for several civil engineering structures where the concrete cost is much higher due to better quality. For instance, in tunnel linings and marine structures wherein the security is a major issue or in structures in which there is limited accessibility for repairing and maintaining. In such special circumstances, even if the self-healing agent-incorporated concretes cost higher, it should not be too burdensome looking at the safety and future benefits.

11.5   ECONOMY OF NANOMATERIAL-BASED CONCRETES

Generally, concrete is the distinct construction material exploited worldwide. It has been documented that over 2.6 billion tonnes of OPC were manufactured in the year 2007 [21] worldwide, amounting to above 17 billion tonnes. OPC was used in varieties of products such as building basement, walls, footpaths, lamp posts, bridges, dams, tall towers, and skyscrapers to cite a few. Usually, concrete goods are proposed to achieve long lifetime and tolerant against local aggressive atmospheric conditions. Eventually, these concrete structures are mostly demolished and recycled upon reaching the final stage of their service life. Besides, the building sector is such that it is easy to apply process innovations instead of modernizations of disruptive goods. Manufacturing combines the products from varieties of supplies and skilfulness in a

FIGURE 11.5   The C–S–H clusters: (a) TEM image and (b) molecular model generated [20].

broad array of trading into a solitary completed structure. An alteration in the built structure can be evaluated by the construction corporation itself. Moreover, a considerably novel product produced by the supplier needs understanding and approval by the architects, engineers, and the client before being applied by the skilled and trained on-site workers.

Several factors must be accounted for to develop nanotechnology-enrouted concretes. First, concrete and related products must be manufactured at large scale. Even if the cost of expensive concrete structures becomes lower, it must be capable of handling massive material in a safer and environment-friendly way. Second, innovations require to be methodically developed with field testing to achieve the understanding and assurance in the construction sector. Lastly, structures of concrete are hard to destroy and need explosives or high energies for breaking up. Thus, nanotechnology-based concrete production should be compatible to these conventional practices. With these constrictions, the early nanomaterial implementations in self-healing purposes must render notable benefits in terms of extra functions with comparatively low quantities of nanomaterials. This low amount can be offered via standard construction practices and must not influence the materials' performance. Innovative products (smart self-healing concrete) must be able to advance the delivery of the traditional materials including the control of released admixtures to likely penetrating the marketplace.

11.6   ENVIRONMENTAL SUITABILITY OF NANOMATERIAL-BASED CONCRETES

Commonly, gaseous CO_2 is released from OPC concrete during the cement clinker's de-carbonation of lime and calcination reactions. Using nanomaterial-based self-healing technology, emission of CO_2 can remarkably be reduced. Currently, world's exclusive development of concrete is beyond imagination [22]. History tells us that without concrete, the wonderful structures such as Sydney Opera House, the Chrysler Building, or Taj Mahal could not have existed. Furthermore, the skyscrapers in metropolis all over the world would have reached to such striking altitudes without the usage of concretes. In every aspect, durability of concrete played a remarkable role to erect those historic buildings centuries ago without modern technology and qualified engineers. Concrete in its own right is an integrated part of everyday life. Briefly, the manufacturing of smart and new concrete demands much money where billion tonnes of raw materials are wasted annually because of its inefficient production processes. Moreover, the production of OPC (primary concrete binder) adds to over 5% of the total released greenhouse gases annually worldwide. It enforces threat to our environment where world development is striving for sustainable and green building deployment [23]. Therefore, the future aim is targeted to build cleaner, safer, efficient, reliable, and stronger smart material alternative to concrete than the conventional OPC-based concretes. In this spirit, the notion of nanomaterial-based smart concrete and self-healing technology has been coined.

REFERENCES

1. Diab, A.M., et al., Effect of nanomaterials additives on performance of concrete resistance against magnesium sulfate and acids. *Construction and Building Materials*, 2019. **210**: p. 210–231.
2. Rathi, V. and C. Modhera, An overview on the influence of nano materials on properties of concrete. *International Journal of Innovative Research in Science Engineering and Technology*, 2014. **3**(2): pp. 9100–9105.
3. Shih, J.-Y., T.-P. Chang, and T.-C. Hsiao, Effect of nanosilica on characterization of Portland cement composite. *Materials Science and Engineering: A*, 2006. **424**(1–2): pp. 266–274.
4. Jo, B.-W., et al., Characteristics of cement mortar with nano-SiO_2 particles. *Construction and Building Materials*, 2007. **21**(6): pp. 1351–1355.
5. Makul, N., Advanced smart concrete – a review of current progress, benefits and challenges. *Journal of Cleaner Production*, 2020: p. 122899.
6. Sbia, L.A., et al., Evaluation of modified-graphite nanomaterials in concrete nanocomposite based on packing density principles. *Construction and Building Materials*, 2015. **76**: pp. 413–422.

7. Jeevanandam, J., et al., Review on nanoparticles and nanostructured materials: history, sources, toxicity and regulations. *Beilstein Journal of Nanotechnology*, 2018. **9**(1): pp. 1050–1074.

8. Samadi, M., et al., Waste ceramic as low cost and eco-friendly materials in the production of sustainable mortars. *Journal of Cleaner Production*, 2020: p. 121825.

9. Huseien, G.F., et al., Development of a sustainable concrete incorporated with effective microorganism and fly ash: characteristics and modeling studies. *Construction and Building Materials*, 2021. **285**: p. 122899.

10. Shah, K.W. and G.F. Huseien, Biomimetic self-healing cementitious construction materials for smart buildings. *Biomimetics*, 2020. **5**(4): p. 47.

11. Hassellöv, M., et al., Nanoparticle analysis and characterization methodologies in environmental risk assessment of engineered nanoparticles. *Ecotoxicology*, 2008. **17**(5): pp. 344–361.

12. Kumar, P. and L. Morawska, Recycling concrete: an undiscovered source of ultrafine particles. *Atmospheric Environment*, 2014. **90**: pp. 51–58.

13. Rana, A., et al., Recycling of dimensional stone waste in concrete: a review. *Journal of Cleaner Production*, 2016. **135**: pp. 312–331.

14. Li, W., et al., Effects of nanoparticle on the dynamic behaviors of recycled aggregate concrete under impact loading. *Materials & Design*, 2016. **112**: pp. 58–66.

15. Chowdhury, T., et al., Is the commercial sector of Bangladesh sustainable? – viewing via an exergetic approach. *Journal of Cleaner Production*, 2019. **228**: pp. 544–556.

16. Chowdhury, H., et al., A study on exergetic efficiency vis-à-vis sustainability of industrial sector in Bangladesh. *Journal of Cleaner Production*, 2019. **231**: pp. 297–306.

17. Huseien, G.F., et al., Evaluation of alkali-activated mortars containing high volume waste ceramic powder and fly ash replacing GBFS. *Construction and Building Materials*, 2019. **210**: pp. 78–92.

18. McLellan, B.C., et al., Costs and carbon emissions for geopolymer pastes in comparison to ordinary Portland cement. *Journal of cleaner production*, 2011. **19**(9–10): pp. 1080–1090.

19. Hamers, R.J., Nanomaterials and global sustainability. *Accounts of Chemical Research*, 2017. **50**(3): pp. 633–637.

20. Pacheco-Torgal, F. and S. Jalali, Nanotechnology: advantages and drawbacks in the field of construction and building materials. *Construction and Building Materials*, 2011. **25**(2): pp. 582–590.

21. Raki, L., et al., Cement and concrete nanoscience and nanotechnology. *Materials*, 2010. **3**(2): pp. 918–942.

22. Bondar, D., Alkali activation of Iranian natural pozzolans for producing geopolymer cement and concrete, 2009, The University of Sheffield.

23. Kambic, M. and J. Hammaker, Geopolymer concrete: the future of green building materials, 2012: pp. 1–7.

Appendix
Questions and Answers

Question 1: What is nanotechnology?
Answer: The definition most frequently used by government and industry involves structures, devices, and systems having novel properties and functions due to the arrangement of their atoms on the 1–100 nm scale. Many fields of endeavour contribute to nanotechnology, including molecular physics, materials science, chemistry, biology, computer science, electrical engineering, and mechanical engineering.

Question 2: What is nanotechnology considered to be?
Answer: Nanotechnology can be defined as the manipulation of shape and structure of materials at the nanoscale in order to design, characterize, and produce useful structures, devices, and systems. The nanoscale refers to the objects between 1 and 100 nm in size, where 1 nm is equal to 1×10^{-9} m.

Question 3: What are nanomaterials?
Answer: According to the definition of the International Organization for Standardization (ISO), manufactured nanomaterials of organic or inorganic origin are differentiated into three types of nano-objects which are smaller than 100 nm in at least one dimension:

  i. Spherical structures (e.g. nanoparticles and fullerenes)
  ii. Fibrous structures (e.g. nanotubes)
  iii. Extremely thin layers (e.g. nanoplatelets)

Question 4: What are nano core-shell or nanocapsules?
Answer: Organic compounds such as liposomes, micelles, and vesicles are added to foods to encapsulate other substances such as vitamins or flavourings, transport them through the body, and release them at exactly the right spot. As the size of these"transport containers" is often in the nanometre range, they are also referred to as nanocapsules.

Question 5: Are nanomaterials used in the construction industry?
Answer: It is being reported that nanomaterials are used as enhancement and additives in concrete, steel, and paints. In the construction industry, one of the possible solutions for a sustainable future is to introduce novel technologies to improve the durability of materials and increase the life span. Presently, nanotechnology creates new possibilities to control and improve material properties for civil infrastructures. Nanotechnology refers to the sub-atomic manipulation of materials which can be achieved by using a combination of engineering, chemical, and biological approaches.

Question 6: Why develop nanotechnology?

Answer: Gaining better control over the structure of matter has been a primary project of our species since we started chipping flint. The quality of all human-made goods depends on the arrangement of their atoms. The cost of our products depends on how difficult it is for us to get the atoms and molecules to connect up the way we want them. The amount of energy used – and pollution created – depends on the methods we use to place and connect the molecules into a given product. The goal of nanotechnology is to improve our control over how we build things, so that our products can be of the highest quality, while causing the lowest environmental impact. Nanotech is even expected to help us heal the damage our past cruder and dirtier technologies have caused to the biosphere.

Nanotechnology has been identified as essential in solving many of the problems facing humanity. Specifically, it is the key to addressing the Foresight Nanotech Challenges:

1. Providing Renewable Clean Energy
2. Supplying Clean Water Globally
3. Improving Health and Longevity
4. Healing and Preserving the Environment
5. Making Information Technology Available to All
6. Enabling Space Development

Question 7: How can nanotechnology promise to build products with both extreme precisions in structure and environmental cleanliness in the production process?

Answer: Traditional manufacturing builds in a "top down" fashion, taking a chunk of material and removing chunks of it – for example, by grinding, or by dissolving with acids – until the final product part is achieved. The goal of nanotechnology is to instead build in a "bottom-up" fashion, starting with individual molecules and bringing them together to form product parts in which every atom is in a precise, designed location. In comparison with the top-down approach, this method could potentially have much less material left over, greatly reducing pollution.

In practice, both top-down and bottom-up methods are useful and being actively pursued at the nanoscale. However, the ultimate goal of building products with atomic precision will require a bottom-up approach.

Question 8: Where is nanotechnology being developed?

Answer: Research and development of nanotechnology is taking place worldwide. As this is written, government spending is at approximately US$1 billion in each of four global areas: (i) the United States, (ii) Europe, (iii) Japan, and (iv) the rest of the world, including China, Israel, Taiwan, Singapore, South Korea, and India. Similar amounts are said to be being spent in the private sector, with these figures being quite difficult to determine accurately due to the breadth of the nanotech definition, which includes a large number of older technologies.

Question 9: Are there any safety or environmental issues with the nanotechnologies in use today?

Answer: Concerns have been raised regarding potential health and environmental effects of the passive nanostructures termed"nanoparticles". Regulatory agencies and standard bodies are beginning to look at these issues, though significantly more funding for these efforts is required. Foresight is working with the International Council on Nanotechnology to address these concerns.

Question 10: What are nanoparticles, nanotubes, and nanofilms?

Answer: Depending on the shape, the application, or the components, nanomaterials may be called by a variety of different names, including nanoparticles, nanotubes, nanofilms, nanoshells, nanospheres, nanowires, nanoclays, nanoconcrete, nanopolymers, and much more. Other nanomaterials have distinct qualities that have led researchers to call them by other non-nano prefix names such as quantum dots or graphene. Generally speaking, nanomaterials are objects with one or more dimension at the nanoscale. Efforts to standardize these words are currently underway, for example, by the ISO.

Index

acid 28, 35, 50, 73, 86–96, 125
activation 49, 123, 170
aggregates 38, 50, 117–120, 145, 195
aggressive 49–50, 73, 81, 201
alkali-activated 25, 49–50, 86, 92, 197
antivirus 159–160

binder 14, 22–25, 50–56, 66, 86, 137, 199
biological 4, 138–139, 159, 162, 183, 192
building 1, 6, 16, 36, 45, 99–107, 125–127
bulk density 22, 53–55

calcium 13, 22–27, 40, 49, 67, 85, 159
carbon dioxide 21, 39, 44, 67, 138, 199
carbon nanotube 6, 13, 22, 51, 121, 145
cement 6, 15, 21–28, 35, 73, 119, 120
clinker 14, 21, 203
coarse aggregates 195
compressive strength 16, 24–26, 41–55, 60
concrete 4–6, 13–16, 21–29, 35–42, 77–80
construction 1, 6, 11, 13, 16, 43, 107
cost 13, 22, 36, 74, 100, 102–107, 138
crack 6, 13, 26–29, 35–43, 84, 92

deterioration 28, 35, 39, 84–92, 139, 141
development 44, 94, 108, 128
ductility 15, 145
durability 13–16, 27–30, 38, 44, 65, 82

eco-friendly 44, 51, 204
economic 36, 68, 73, 109, 125, 139, 196
economically 36, 67
elevated temperatures 49, 68, 74, 90–96
energy 1, 4, 6, 12, 21–30, 50, 73, 99–108
environment 2, 14, 17, 35–44, 73, 125, 161

fine aggregates 14, 92, 198
flexural strength 13, 25, 38, 51, 69, 146
fly ash 13, 14, 17, 24, 28, 62, 197
fresh properties 23, 70, 118

gels 40, 50, 61, 65, 73, 77, 96
geopolymer 61, 65–66
greenhouse 21, 44, 49, 99, 137, 203
gypsum 14, 21, 29, 88, 96, 103, 124–126

hazardous 1, 49, 193
hydration 6, 13, 22–28, 38, 40–43, 53, 66

improvement 1, 11, 22, 24–26, 62
infrastructure 21, 35, 51, 139, 141, 187

landfill 49, 184

mechanical properties 4, 6, 13, 55, 66, 127
micro 11, 16, 28
microstructure 16, 22–29, 40, 57, 86, 145
mineral 14, 39, 41, 65
modulus 57, 147
mortar 23–29, 38, 49–58, 75

nanofiber 8, 32, 133
nanomaterials 1–7, 11–15, 23–44, 85–97
nanoparticles 1–7, 14, 23, 122, 147–150
nanoscience 2–4, 138, 145, 151, 162, 196
nano silica 15, 22, 38–40, 51, 66, 88, 144
nano sized 6, 167, 181, 191
nanotechnology 1–7, 11–15, 22–38, 100
natural 4, 7, 12, 103–110, 118–122

optimum 26, 24, 30, 55, 68, 82, 115

particle size 1, 36, 38–39, 113–115, 146
permeability 14, 16, 41, 51, 147–149
phase change materials 99, 101,
    105–117
pozzolanic 22–26, 38, 50–52,
    140–151, 193

quartz 42, 61, 88, 93, 120

reduction 16, 22, 28, 61, 66, 77, 80
replacement 2, 24, 39, 51, 62, 66, 75–77
resistance 3, 11, 14, 24, 79–88, 96

safety 36, 43–50, 140, 162, 171, 181–183
silica 6–7, 13–16, 23–29, 38, 40–42,
    50, 56
sodium 29, 49, 68, 115, 142, 198
sustainability 22, 35, 44, 65, 126, 137, 195

titanium 13–15, 24, 41, 125, 146, 169, 182

viscosity 51–53, 122, 148

workability 23, 39, 50–53, 67,
    195–196

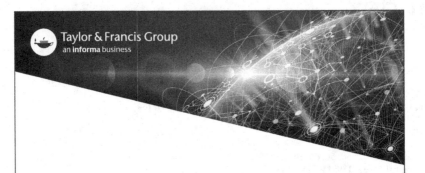

Printed in the United States
by Baker & Taylor Publisher Services